Sponges

Tedanya (Tedanya) anhelans, from Portsea Pier, Port Phillip Bay. Julian Finn

A Museum Victoria
Field Guide to Marine Life

Sponges

Lisa Goudie
Mark Norman
Julian Finn

MUSEUMVICTORIA

Series editors
Mark D. Norman
Melanie Mackenzie

Other titles in the series

An introduction to marine life
Barnacles
Crabs, hermit crabs and allies
Shrimps, prawns and lobsters

Published by
Museum Victoria 2013

Museum Victoria
Publishing
GPO Box 666
Melbourne Vic 3001 Australia
Tel + 61 3 8341 7370 or 8341 7536
publications@museum.vic.gov.au

Museum Victoria
www.museumvictoria.com.au

PRINTED BY
BPA Print Group
ORIGINAL DESIGN BY
Propellant
TYPESET BY
the printer's drawers

National Library of Australia Cataloguing-in-Publication entry
Goudie, Lisa.
Sponges / Lisa Goudie, Mark D. Norman and Julian Finn.

Includes bibliographical references and index.
Sponges--Australia, Southeastern.
Other Authors/Contributors:
Norman, Mark Douglas.
Finn, Julian.
Museum Victoria.
ISBN: 9780980381399 (pbk.)
9781921833151 (e-pub)
9781921833182 (e-pdf)

593.40994

The content of this publication is provided for information purposes only and is not intended as a substitute for professional advice. Museum Victoria does not accept liability for any damage, loss, injury of death to any person in relation to the information or advice (or the use of such information or advice) which is provided in this publication.

Funding for the Museum Victoria Field Guides to Marine Life is gratefully acknowledged from the Australian Government through a grant from its Natural Heritage Trust. The grant was facilitated by the Port Phillip and Westernport Catchment Management Authority. Publication of the series is funded in part by the Victorian Coastal Council and Parks Victoria.

FRONT COVER IMAGE
Sycon sp.
Photographer: Mark Norman

IMAGES ON BACK COVER
Dactylia sp. LG1
Speciospongia purpurea
Echinoclathria sp. LG1
Photographer: Mark Norman

MUSEUM VICTORIA FIELD GUIDES TO MARINE LIFE

These field guides to marine life enable the amateur naturalist, beachcomber or environmental scientist to identify the marine animals most commonly found on the shore or in shallow waters along the coast of the state of Victoria, Australia. South-eastern Australia is characterised by a rich marine fauna, with many species found nowhere else. Commonly, species found along the Victorian coast also occur in Tasmania, southern New South Wales, and along the southern coast of the continent, through South Australia, into southern Western Australia.

This series aims to cover the common marine animals, and each book deals with a different group. More species live on the Victorian shore and in its shallow waters than are included in each book, and many more inhabit the deeper waters of Bass Strait and beyond. See **Further Information** at the end of this guide, and Museum Victoria's website: *museumvictoria.com.au*.

Museum Victoria encourages individuals to explore the diversity of coastal habitats, but discourages unnecessary removal of specimens from their natural environment. Museum scientists are interested in new discoveries, and unusual findings can be reported to the Discovery Centre at Museum Victoria, or to Reef Watch Victoria: *info@reefwatchvic.asn.au*.

Reef Watch Victoria is a community-based marine monitoring program for Victoria's temperate marine environment. Divers and snorkellers conduct regular surveys at their favourite Victorian reef sites using the Reef Watch monitoring kit. For more information, go to: *www.reefwatchvic.asn.au*.

The Marine Research Group (MRG) is a branch of the Field Naturalists Club of Victoria which has had a long and productive working relationship with Museum Victoria, actively undertaking research into Victoria's rich marine life, and providing curatorial and survey assistance with Museum Victoria's extensive marine invertebrate collection. The MRG meets regularly, welcomes new members, and can be contacted at: *www.fncv.org.au*.

A large proportion of the images and information presented in this guide has resulted from the 'Under the Lens' partnership between Museum Victoria and Parks Victoria. Along with ports and harbours, Parks Victoria manages marine national parks and sanctuaries throughout Victoria. For more information, go to: *parkweb.vic.gov.au*.

CONTENTS

Chondropsis cf kirki. Mark Norman

What are sponges?

Most people think of sponges as something they use in the kitchen or bathroom. Although now more commonly made of synthetic foam, these household objects were historically made up of the skeletal remains of once-living true sponges. The term sponge means 'to squeeze' but not all sponges are spongy. In fact they can be soft and slimy, fibrous, prickly, sandy and crumbly, or 'as tough as old boots'.

Bathroom sponge. Lisa Goudie

Sponges are one of the oldest and simplest life forms on earth. Until the 18th century they were classified as zoophytes ('plant-animals'). However, sponges are animals belonging to a group known as the Phylum Porifera, or literally, 'pore-bearers'. Sponges are simple, multicellular organisms made up of different cell types, each with a different function. Unlike all other animals, however, the cells of sponges are not arranged or grouped to form tissues or organs. Sponges are not colonies of individual animals but rather collections of individual cells, one type of which forms a continuous outer, skin-like layer. The vast majority of sponges are filter feeders attached to the sea floor, drawing seawater through their bodies and with it the food and oxygen they need to survive. In doing so, they provide an important link between the water mass and the seafloor. Sponges are united as a group of animals by the possession of a unique cell type – the **choanocyte** or collar cell (see ***Sponge internal structure***). Choanocytes each have a central flagellum, or filament, that beats like a tadpole's tail, drawing the water in. This current draws food particles in with it, which are trapped by the surrounding collar of small hair-like structures called cilia.

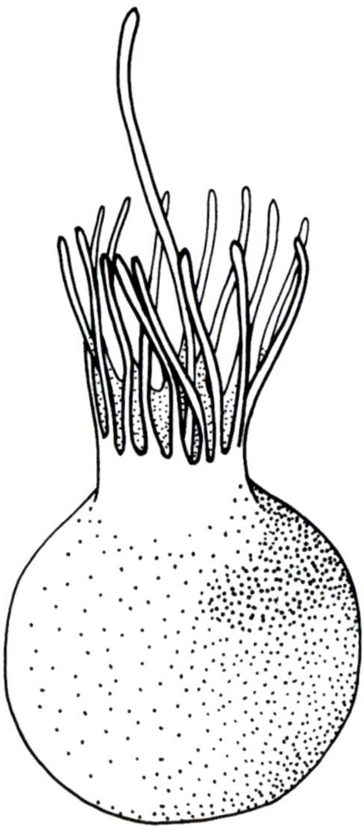

Choanocyte cells produce the current flow through sponges.

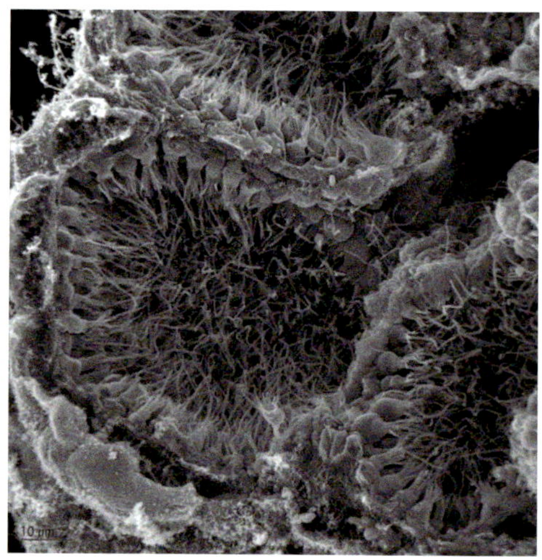

SEM showing sponge choanocytes.

Distribution and habitats

Of the thousands of sponge species in the world, the vast majority are found in marine waters. Only a fraction of these, approximately 150 species, live in freshwater systems. Sponges live at all depths of the ocean from intertidal habitats to depths of up to seven kilometres. Sponges are found from the warm waters of the tropics, where they are important reef-building organisms, to the cold oceans at the poles where glass sponges are a dominant component of the **benthic** (bottom-dwelling) fauna.

A deep-sea stony sponge from eastern Australia (left) and a freshwater sponge from Lake Condah region in Western Victoria. Mark Norman

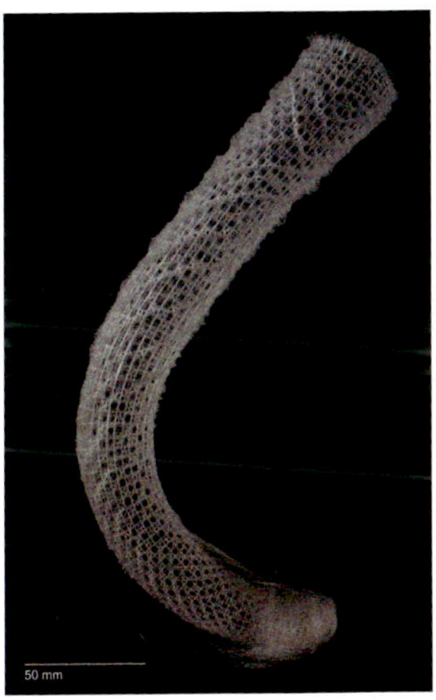

A glass sponge from the Museum Victoria collection. David Paul

Glass sponge reefs recently discovered off the west coast of North America at depths of over 200 m. Image courtesy of Sally Leys

Sponges typically grow attached to hard surfaces, so are common in rock and coral reef habitats. Sponges also occur in soft sediment substrates, where they play an important role in consolidating the sand, shell grit or mud to form anchors for their own growth as well as for a host of other organisms found in these habitats.

Some sponges even live on the surfaces of other animals, either by settling directly on them during their earliest life stages, as **larvae**, or by being placed there by the host animal (see ***Biotic associations***).

How many species?

The number of sponge species currently documented worldwide is about 8500. Estimates of the total number of species, including those yet to be discovered and described, suggest at least twice this figure. Of the thousands of species thought to occur in Australian waters, about 1500 have been officially documented, 1000 of these being recorded in southern Australian waters.

There are three classes of sponges. By far the most commonly occurring are those in the Class Demospongiae. Sponges in this class are mostly marine and together they exhibit a great variety of colours and growth forms. While representatives of the class inhabit all oceans and some freshwater systems, they also demonstrate a high degree of **endemicity** (localisation).

Evolution and history

Sponges are not only the simplest of animals but they are also among the oldest. Dating of fossil sponges has shown us that they have lived in our oceans for around 600 million years. Fossil evidence alone, however, does not provide a clear picture of the path of early animal evolution. Recent genetic studies on the molecules fundamental to all cells have shown that an ancient sponge-like ancestor was in fact the first true animal. It is quite likely that sponges share this ancestor with the living, unicellular organisms known as **choanoflagellates**. Choanoflagellates are almost identical in both shape and function to the choanocyte cells of sponges. Both share a collar of closely packed hair-like cilia surrounding a single, beating, whip-like flagellum – a very successful evolutionary unit.

In ancient seas, sponges were the first reef builders, their skeletons forming reefs that would rival the coral reefs of today. In the warm, shallow waters of the Tethys Sea 160 million years ago, when sponges were very diverse, one such reef occupied most of what is now Europe. Fossil remains of the reef are exposed today in rocks over an area spanning from central Spain to eastern Romania. These early sponges had cemented calcium carbonate bases. It is possible that ocean acidification, similar to that being discussed today in association with climate change, may have wiped out these sponge reefs in past mass extinction events.

The sponges share ancestry with the planktonic single-celled organisms known as choanoflagellates. Kerry B. Clark

Fossil sponge reefs exposed in southern Poland. Courtesy of A. Pisera

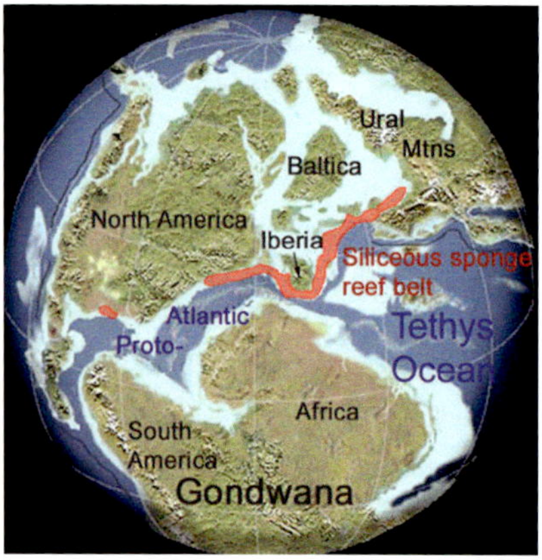

Extent of fossil sponge reefs of Europe.
Manfred Krautter following Ron Blakey

Recent exploration of deep-sea environments and less accessible habitats has revealed sponge groups previously thought to be extinct. In the early 1980s, divers exploring at the back of deep caves in the Coral Sea off the north-eastern coast of Australia discovered sponges thought to have been extinct since the end of the Cretaceous period, 65 million years ago. *Vaceletia*, a **genus** of these recently described sponges, has now been discovered widely in the Indo-Pacific region, both in caves and in deep water, growing as worm-like tubes up to 30 centimetres in diameter. These rock-hard sponges possess concentric growth rings, much like those seen in trees, which suggest individual sponges may have been living for about 5000 years.

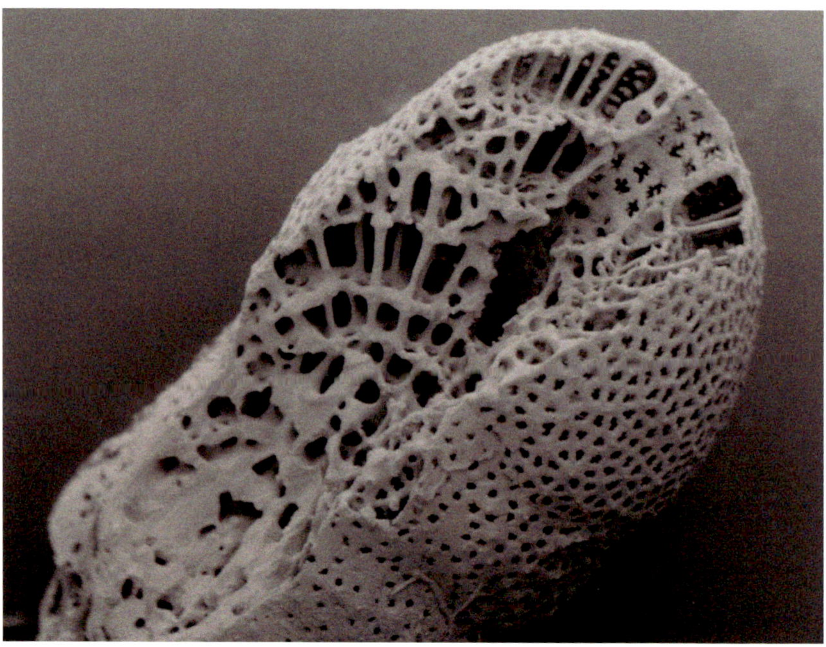

The living fossil sponge genus, *Vaceletia*. Jean Vacelet

Body form

As a group, sponges exhibit a huge range of sizes, colours, textures and growth forms. Larger free-standing sponges may take the form of volcano-shaped mounds, spheres, vases, blades, whips, fingers or branched plant-like structures. Some are encrusting, forming a thin layer over hard surfaces such as rock, coral, shell or the backs of crabs and scallop shells. Other sponges live buried below soft substrates, raising **fistules** (finger-like projections) above the sediment to access the seawater. Certain sponges use chemicals to bore into coral, rock or shell to make safe tubes in which to live. Sponges are much more abundant than they first appear as many species are hidden under the substrate, inside rocks or in crevices.

An example of some sponge forms shown in this guide.

The sponge body plan is organised around a system of internal water canals and chambers through which flows a unidirectional current of water. This current is generated by the beating of the fine filaments of choanocytes lining the canals and chambers. Water containing nutrients and oxygen is drawn in through the many **ostia** (small **inhalant** pores) and is expelled, along with waste and reproductive products, out through the larger **oscules (exhalant** openings). Ideally the arrangement of these inlets and outlets is such that they are kept apart to prevent contamination of the incoming seawater.

The simplest body structure found in sponges is termed **asconoid**. Sponges with this structure are small and spherical with a single exhalant opening. This simple system limits the maximum size attainable by the sponge. Any increase in size in sponges necessitates a more complex internal structure. This is achieved by folding of the body wall to form canals, resulting in an increase in the surface area covered by choanocyte cells that are in contact with the water current. This enables an increase in the volume of water being pumped through the sponge body. Sponges exhibiting some folding of the body wall have a **syconoid** structure. Further folding of the body wall is seen in the more highly evolved structure of **leuconoid** sponges. The canals of these sponges have undergone further folding to form chambers, enabling them to pump up to 10 times their own volume each hour. These sponges can attain a considerable size and generally have many exhalant openings. Leuconoid sponges exhibit a wide range of growth forms from thinly encrusting to erect and branching, or **massive** (large and without a definable shape).

Being filter-feeders, sponges thrive in areas of strong current where the levels of food particles in the water column are high. The environmental conditions greatly influence the growth form of many sponges. In areas of high current, a sponge species might take a low encrusting form while the same species in areas of less water flow might grow as a more erect and branching form. This **plasticity** of form presents one of the biggest difficulties in identifying sponges. For this reason shape alone cannot be relied upon for classification but must be used in conjunction with a number of other features (see ***Classification of sponges***).

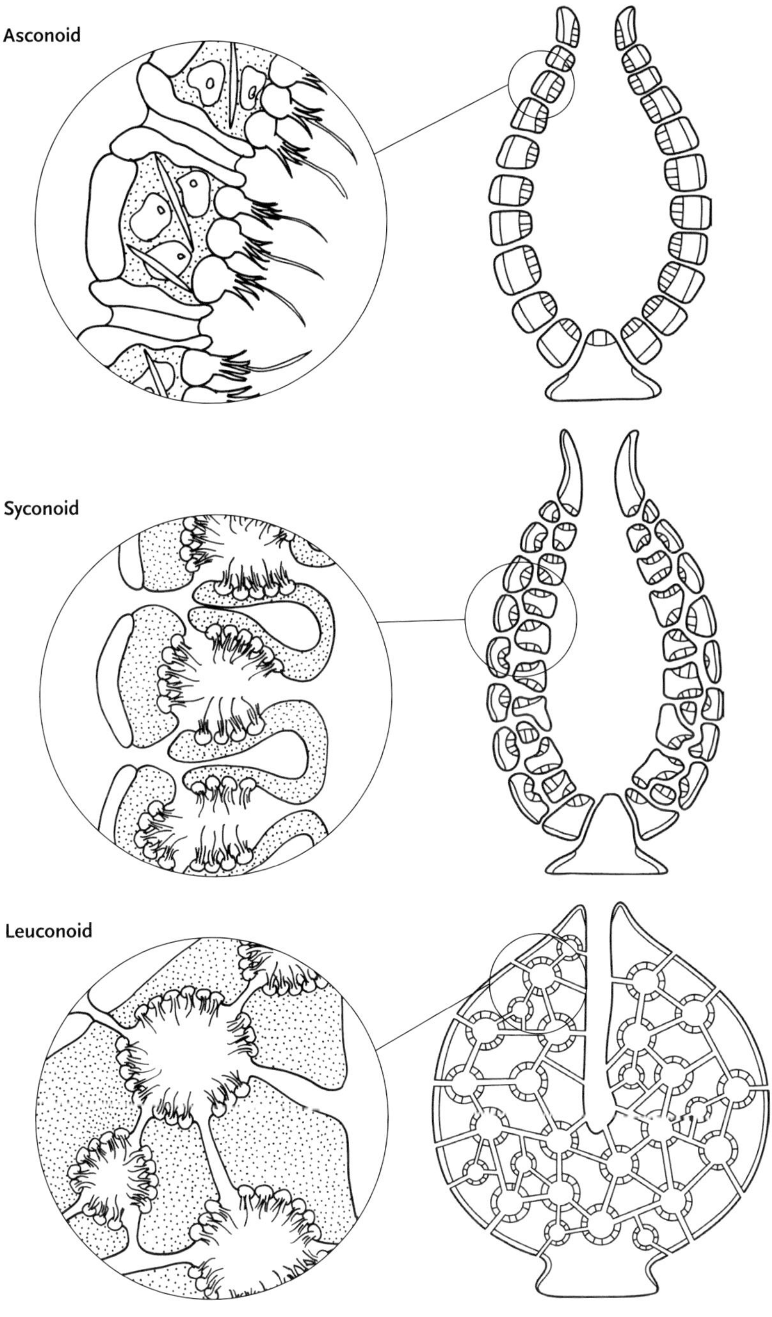

Three stages of sponge body forms from simplest (asconoid) to the more complex (leuconoid).

Sponge internal structure

Body layers

Most sponges have three body layers:

1. The outer encompassing layer or 'skin' is known as the **pinacoderm**, formed by a single layer of cells called **pinacocytes**.
2. The middle layer or **mesohyl** consists of a **collagen** matrix and the skeletal material. These skeletal components can include **organic spongin** fibres and/or **inorganic spicules** (glass-like microscopic structures made of silica or calcium carbonate).
3. The inner layer or **choanoderm** is the layer of feeding cells, or choanocytes, lining the canals and chambers and is responsible for generating the water current.

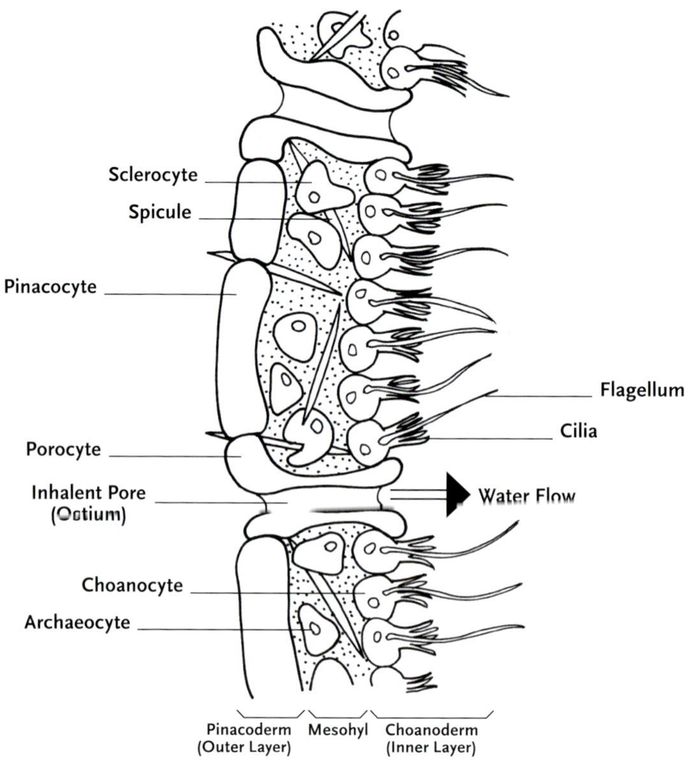

Cross section of a sponge body wall showing the three layers and cell types.

Cell types

Sponges have a relatively small number of cell types, each performing a different function such as current generation, digestion, reproduction, spicule formation or waste removal. The most important cell type is the **archaeocyte**. These highly mobile cells are described as **totipotent**, meaning they are capable of moving about and transforming into other cell types as required. They can be likened to stem cells in humans. Archaeocytes play a critical role in the ability of some sponges to rebuild themselves from fragments. Some sponges can even be passed through a fine silk mesh and are able to reform themselves into a single sponge again on the other side. The high level of motility and plasticity of sponge cells is unique amongst multi-cellular organisms. By contrast, the feeding cells (choanocytes) are fixed once formed.

Sponges do not typically exhibit any movement. However, some special sponge cells called **actinocytes** can contract sufficiently to close ostia as a way of regulating flow and to help prevent silting up. Recent studies of a species of the 'golf ball' sponge, T*ethya wilhelma*, have shown an extraordinary ability of the sponge to move like a 'rolling stone' by rhythmic contractions in the outer, **cortical** layer when conditions become unfavourable at a given location.

Skeleton

The differing growth forms and textures of sponges are derived in part from the nature of the supporting framework or skeleton of each individual sponge. The skeleton may be **collagenous** (having flexible fibres), **spiculose**, a combination of both, or completely lacking. Some sponges incorporate shell fragments, sediment from the seafloor and even spicules shed by other sponges, into their skeletons.

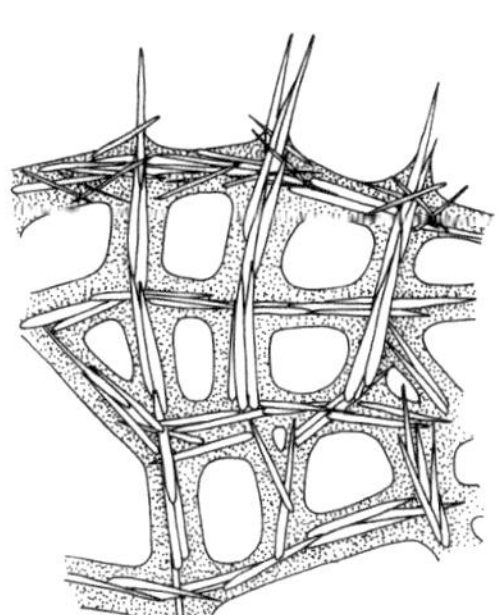

Spiculose skeleton

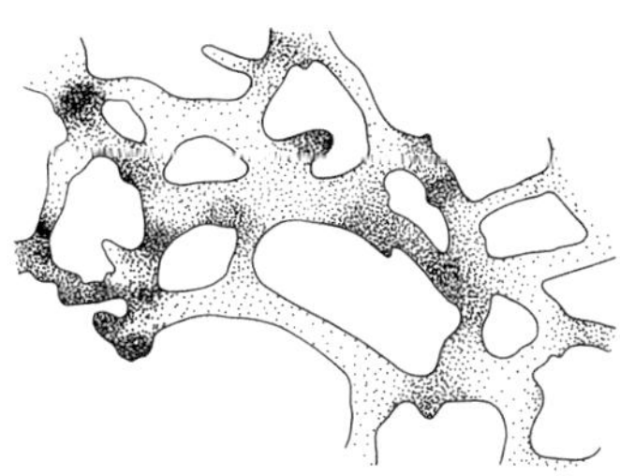

Collagenous skeleton

Spicules

The mineral skeleton of sponges is made up of microscopic structures known as spicules. These are inorganic structures that are secreted by the **sclerocyte** cells and may consist of either silicon dioxide or calcium carbonate. They are produced in distinctive forms to provide specific structural or defensive properties. In the newly discovered but rare carnivorous sponges, special spicules are used to capture animal prey (see ***Carnivorous sponges***).

Knowledge of the spicule type and shape helps to separate sponges into one of three major groupings or classes; the **demosponges**, the **hexactinellids** (glass sponges), and the **calcareous sponges**. Spicules are also used to distinguish all other systematic groupings from order down to species. Spicule **morphologies** (structural forms) are usually characteristic of a given sponge and, once determined under a microscope, they aid sponge identification.

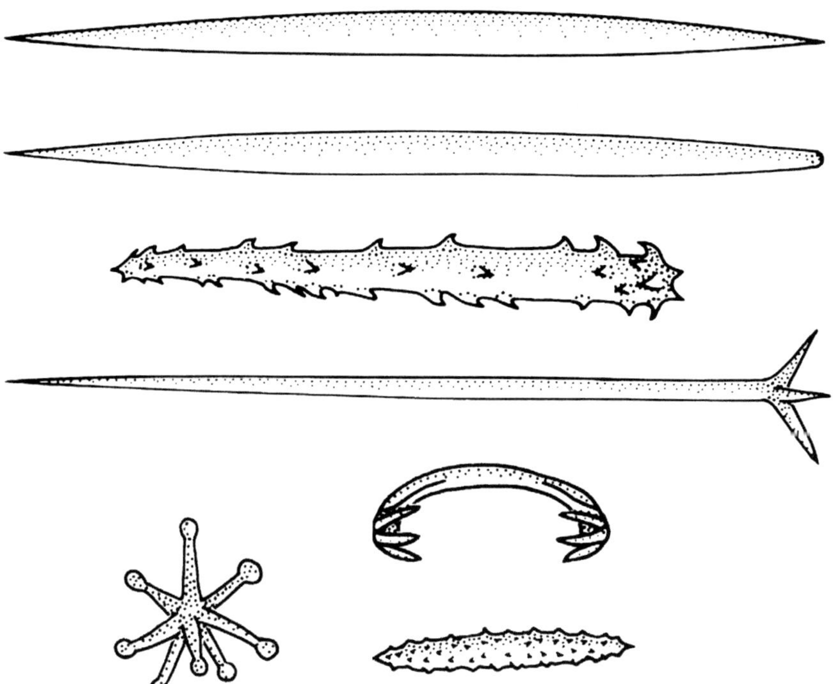

Some of the variety of spicule shapes that form the skeletons of sponges.

Reproduction

All sponges are capable of sexual reproduction – formation of a new organism through the production of egg and sperm cells. Most sponges are **sequential hermaphrodites**, producing egg and sperm cells at different times in order to allow cross-fertilisation with other sponges of the same species. Fertilisation of the egg cell by the sperm cell may take place in the water column but more commonly occurs within the adult sponge itself. The resulting larvae are either free-swimming (and may remain in the water column for 2 to 72 hours), or crawl over the substrate before settling and adhering to a hard surface and growing into an adult sponge. The eggs and sperm may be produced by choanocytes or by archaeocytes. In most sponges, the sperm are captured by the choanocytes and transferred to the egg cells.

Sponges can also reproduce **asexually** through budding or fragmentation, which aids in the dispersal and recruitment of young sponges. The golf ball sponges (genus *Tethya*) are a good example of sponges that use asexual dispersal. These sponges bud tiny reproductive balls on the outside of their body that break loose and roll away to grow into new sponges. Many sponges can reform from a broken fragment dislodged during disturbances such as storms. Some freshwater sponges even have a **gemmule** (resting phase), which can last up to two years in a dried-out spore state.

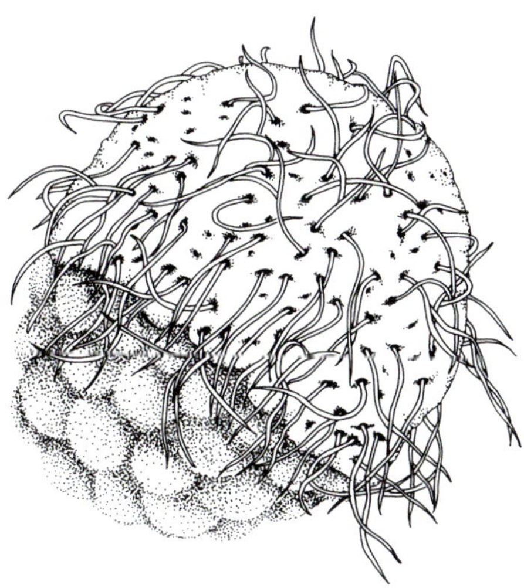

The youngest stage of a sponge is known as an **amphiblastula**.

The golf ball sponges, genus *Tethya*, often bud small asexual **propagules** off their body. Julian Finn

A spawning sponge.
Mark Rosenstein

Age and growth rates

Sponges are generally a long-lived group of animals, many living for decades, some for hundreds, and a few for thousands of years. Others however are short-lived or seasonal, dying back every year. Growth rates also vary between different sponges, possibly in response to changing water temperature. The faster-growing sponges may grow by up to a few centimetres a week. Differing growth rates also appear to be linked to reproductive strategies. Sponges with high growth rates and high reproductive output have increased invasion opportunities in unstable environments, while those with slower growth rates and low reproductive output exhibit good adaptation to more stable conditions, such as in the deep sea.

Studies over many years of the giant barrel sponges, *Xestospongia muta*, in the Caribbean Sea have revealed that individuals up to one metre wide are about 100 years old while those 2.5 metres wide are more than 2000 years old, making them some of the longest-lived animal species in the world.

The Giant Barrel Sponge, *Xestospongia muta*. Dr Joseph Pawlik, UNCW, Wiki

Feeding in sponges

Filter-feeding

Sponges feed by trapping organic particles filtered from the water. They are considered the 'vacuum cleaners' of the ocean, however, current research indicates that sponges are actually more selective in their feeding than previously thought. By drawing water in through the many small ostia distributed over the body surface, each being only a tenth to a hundredth of a millimetre across, they are able to trap and filter organic matter from the seawater including microscopic animals, plants and bacteria. The water travels down canals, then through chambers lined with choanocytes. A choanocyte cell draws the water in using its large beating flagellum and traps food particles with its collar of small hair-like cilia. Other mobile cells in the sponge, such as the archaeocytes, take some of this food and transport it, along with oxygen, to cells throughout the sponge. The filtered seawater, along with waste products, is then returned to the ocean via the large exhalant oscules. In this way the sponges both feed and 'breathe'.

Carnivorous sponges

Carnivorous sponges have only recently been discovered. They belong to a family of predominantly deep-sea sponges, the Cladorhizidae, although one example, *Asbestopluma hypogea*, was found living in a submarine cave off the coast of Provence, France, at a depth of 15 to 25 metres where the water is constantly cold. Lacking the typical canal system of all other sponges, this species captures its prey using filaments covered with hook-like spicules. Acting almost like Velcro, the spicules entrap the hair-like appendages of the unsuspecting prey, tiny

The carnivorous sponge, *Asbestopluma hypogea*, capturing its prey, a mysid shrimp. Jean Vacelet

shrimp-like crustaceans about 10 millimetres long. Other carnivorous species of sponge have since been found elsewhere around the world. These include a deep-sea sponge, *Lollipocladia tiburoni,* observed at a depth of over 2000 metres in the northeast Pacific, growing as a disc on a stalk up to 10 centimetres high.

As these sponges do not possess a true digestive tract, individual cells migrate from throughout the sponge body to concentrate around the prey. Fragments of the prey are digested by archaeocytes and other cells known as **bacteriocytes** in a process that can take 8 to 10 days. The unique feeding habit of this sponge appears to be an adaptation to life in a nutrient-poor environment.

Sponges are constantly competing for space with their neighbours. Mark Norman

Sponge defences

As stationary, bottom-dwelling and slow-growing animals, sponges are extremely vulnerable to three main threats: 1) being eaten, 2) having other creatures settling and growing on them, and 3) competing for space with plants and filter-feeding animals including other sponges. As such they need an arsenal of defences to protect themselves. The skeletal components of their body can provide some defence, particularly in those species with sharp or barbed spicules. However, sponges are best known for their chemical defences, especially in the pharmaceutical world (see ***Sponge use and commercial interest***). They are experts in chemical warfare, producing or containing diverse and largely as yet undiscovered toxins.

It appears that the majority of the toxins found in sponges are either metabolic waste products incidentally produced by the sponge, or toxins that have been modified from these chemicals by the sponge itself. The toxins may be produced to make the sponges unpalatable or even deadly to eat. They can also be used to resist or in fact dissolve neighbours in border disputes. This allows many sponges to compete with faster-growing animal colonies and plants. As a consequence of the dilution factor of the water in which sponges live, these toxins must be extremely potent to be effective. In some cases, the toxic slime oozed by certain sponges can cause severe dermatitis or skin irritation. Care must be taken when handling sponges, even beach-washed or dead ones, as there is a risk of contamination by the toxins and painful puncture wounds from the spicules. Pets on beaches are also at risk, with numerous reports of dogs dying after eating sponges that have washed up on the beach.

Note: First aid
Always seek medical advice for all but minor symptoms. Treatment for marine sponge injuries includes: removal of spicules from the skin with fine forceps where possible, washing of the contact area with soap and freshwater as soon as possible, cold water soaking to relieve pain and reduce local symptoms, and the application of a strong, local steroid cream to the contact area.

Sponges have the ability to vary the production rates of these chemical toxins. When a sponge is under threat, it may increase its production of toxin. Sponge researchers have inserted wedges of other filter-feeding animal colonies into the bodies of sponges and found that the sponges produced 100 times their normal resting concentrations of toxins. This is supported by the fact that the most toxic sponge species come from habitats with high competition for space and access to water currents. However, not all sponges produce toxins. Other strategies are also employed to ensure survival. Studies have shown that chemically undefended sponges, especially those with a more erect growth form, exhibit significantly faster healing rates following damage in order to prevent infection and maintain flow of the feeding current.

While some sponges have been found to be full of toxic, inorganic elements that they extract from seawater, such as zinc and cadmium, most sponge toxins are organic compounds. It is the organic compounds with bioactivity in which pharmaceutical companies are most interested (see ***Sponge use and commercial interest***).

Predators of sponges

Despite their impressive defences, sponges do have a few predators. Many molluscs have evolved their own defences to toxins and are able to feed directly on sponges. Chitons (such as species of the genus *Notoplax*), keyhole limpets (genus *Diodora*) and cowries (such as *Zoila friendii*) use their **radula** (toothed tongue) to cut into and eat sponges. Some molluscs feed on the mucous produced by sponges.

The cowrie, *Zoila friendii*, feeding on a sponge, *Spheciospongia papillosa*. Peter Clarkson

A nudibranch living on a sponge. John Chuk

The most impressive sponge predators are the sea slugs or nudibranchs. Many of these tiny molluscs have no shell for protection. Instead they have evolved a way of safely incorporating sponge toxins into their bodies to use for their own defence. Some can even further modify the toxins to make them stronger, while others such as the Spanish Dancer, genus *Hexabranchus*, pass the toxins into their egg ribbons for protection. These nudibranchs often mimic the colours and forms of the sponges on which they live and feed, as do some polychaete worms.

Many species of crustaceans such as snapping shrimps (Family Alpheidae), amphipods and isopods are parasitic on sponges, both living on and eating them. Larger animals that feed on sponges include sea cucumbers, leatherjackets, boxfishes, pufferfishes and coral reef fishes. These latter include angelfishes, parrotfishes and wrasses. Green, Hawksbill and Leatherback Turtles also feed on sponges. For some species, sponges comprise up to 95 percent of the diet.

A Hawksbill Turtle excavating coral rubble to feed on sponges. Mark Norman

Biotic associations

Sponges interact with other marine life in diverse ways. On soft substrates and in areas of high current flow, sponge gardens provide habitat for many marine species. Individual sponges physically provide substrate or safe habitat for many organisms, particularly algae, feather corals (hydroids), crustaceans and even fishes. Such residents can cause problems so some sponges have evolved grills or sieves across the oscules to prevent animals occupying these tunnels and impeding vital water flow.

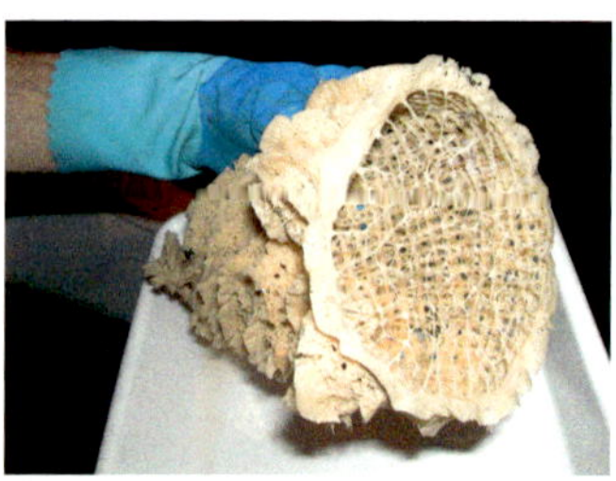

A deep-sea sponge from the Tasman Sea complete with a net grill protecting the vase opening. Mark Norman

Other animals have developed **symbiotic** relationships with sponges. In such relationships the sponge may provide camouflage and/or chemical protection in return for being supported or taken to areas of high water flow and food availability. The bodies of some sea tulips, or ascidians, are cloaked in a tight-fitting layer of living sponge (genus *Darwinella*). The Doughboy Scallop, *Mimachlamys asperrima*, spends its life with a live coating of sponge, commonly a species of *Crella*, on the outsides of its two shells.

Examples of some biotic associations: Sea tulips (top), Doughboy Scallop (above) (Mark Norman), and Great Spider Crab (left) (David Paul)

Many crabs attach pieces of living sponge to their exoskeleton as disguise and/or chemical defence. Spider crabs (genus *Leptomithrax*) and many decorator crabs have small hooks on their skeleton to grip the sponge. As the sponge grows, the crabs must continually trim it to prevent becoming top heavy. Other crabs like the sponge crabs (Family Dromiidae) wear a single tight-fitting cap of sponge held on by their legs. If they become separated from their cap, they appear to do a forward somersault into the hat to put it back in place.

In the late 1990s it was reported that a wild population of bottlenose dolphins in Shark Bay, Western Australia were using sponges as 'tools'. Members of the group were observed breaking cup-shaped sponges of the genus *Echinodictyum* from the sea floor to wear over their rostrum or beak, probably as a protective 'glove' when foraging for food in the sand.

A bottlenose dolphin wearing its protective sponge cup. Amanda Coakes/monkeymiadolphins.org

Not all biotic associations of sponges are with big creatures. Some sponge groups possess microscopic cyanobacteria in their cells. In waters with good light penetration, these bacteria can provide additional nutrition through photosynthesis in much the same way they do for shallow-water corals. Experiments have shown that these sponges cannot survive if the bacteria are removed. These assemblages of bacteria harboured by sponges may also play an important role in the fixation of carbon from the earth's atmosphere – an important consideration with relevance to global warming discussions.

Sponge use and commercial interest

Sponges have been of commercial interest to the cosmetic industry in the form of bath sponges for thousands of years. Diving for sponges in the Mediterranean Sea provided the basis for whole communities of people – their life, culture and economies depending entirely upon the sponge fishery. In early times and in shallow waters, a harpoon was used for the collection of sponges, which were located using a glass-bottomed viewing bucket. In deeper waters, sponges were collected by breath-hold (free) diving. Some reports suggest these divers were able to reach depths of 75 metres on a single breath held for two to three minutes. A **bellstone** was often used in early times – this was a flat, marble stone held at an angle to send a diver more quickly to greater depths. But in the 1860s a revolution occurred in the sponge diving industry with the introduction of the diving suit or **scaphander** (called 'Satan's machine' by divers of the day). The suit was made of rubber with a heavy bronze collar and helmet. A rubber hose through which air was pumped connected the suit to the boat, much like the Hookah system used in diving today. This enabled divers to go deeper and to spend more time in the search for diminishing supplies of sponges. At that time, however, the risks of compressed air use and the need for decompression were not known. Tragically one diver in three was left either dead or crippled by decompression sickness, otherwise known as 'the bends'. Enormous demand for natural sponges in Europe led to an expanded search for sponges. Sponge beds were discovered early in the 20th century in Tarpon Springs (off the coast of Florida, USA), and the sponge fishing industry continues there to the present day.

Sponge diver wearing scaphander, Tarpon Springs, Florida, USA. Library of Congress

Aside from their use in the cosmetic industry, natural sponges have been used for a variety of other reasons. In the past they were used by Roman soldiers to line their metal-plated armour. Today they are used in the restoration of art work and in the production of porcelain, leather and wood.

Sponges of the genera *Spongia* and *Hippospongia,* both suitable for use as bath sponges, have been discovered in the waters of northern Australia but do not exist in commercially sustainable quantities. Due to their highly absorbent skeletons there is still a strong global demand for natural sponges. To help meet this demand, researchers and indigenous communities in northern Australia, in conjunction with the Australian federal government, are cooperating to establish sponge farming as an industry. Pilot projects are now operating in Torres Strait and the Palm Island group in Queensland. Another research facility in Western Australia is looking to the aquaculture of sponges to produce chemicals for pharmaceutical research.

It is this expansion into bioprospecting and pharmaceutical research that has more recently seen sponges in demand as a potential source of marine natural products. It is thought that much of the novel chemistry found in sponges is produced by the symbiotic bacteria they harbour, which in some sponges can represent up to 40% of the living tissue mass. Extracts of these organic chemical compounds can be screened for a range of medicinal, agricultural and environmental properties. Research of this type may lead to the development of new drugs to aid human or animal health, agrochemicals to benefit crop production, or organic antifouling agents to deter growth of marine life on ships' hulls. Compounds discovered in some sponges have produced promising leads in the research of many areas of medicine including possible uses as antibiotic, antifungal and anticancer treatments. Most recently they have shown promise against tropical protozoan parasites that cause malaria and leishmaniasis.

The chemicals extracted from wild sponges are the targets of considerable pharmaceutical and bioprospecting interest.
Professor Rob Capon, University of Queensland

Conservation and sponges

Due to their sedentary lifestyle and relatively slow growth, sponges are vulnerable to habitat disturbance. Throughout Australian waters, and elsewhere in the world, they have been heavily impacted by commercial fisheries that employ bottom trawling. In some fisheries, discarded bycatch, including sponges and other **invertebrates**, accounts for more than 90 percent of the catch. Marine protected areas, and calls for a total ban of bottom trawling, aim to protect benthic communities including sponges.

Sponges and other benthic organisms make up to 90% of fishing bycatch. Peter M. Kyne

As for all marine life, climate change and ocean acidification will impact sponges, particularly those with calcareous skeletons. A global study of disease in sponges over the last 100 years has revealed periods of extensive mortality in a number of locations, most notably in populations of commercial bath sponges in the Mediterranean and the Caribbean Seas. Periodic increase in water temperature, associated rising salinity, and stagnant water are thought to have been the likely causes. Though rarely identified, it is thought that the agents responsible for disease in sponges are either bacteria or viruses. Recent studies suggest that these diseases will be more common in the warming future but that, in some cases at least, sponges do have an immune response and are able to recover.

Classification of sponges

Families, genera, species

All living things (bacteria, fungi, plants and animals) are formally classified within a hierarchical system. In the animal kingdom there are many major groups (phyla). Human beings for example belong to the phylum Vertebrata, along with birds, reptiles, amphibians and fish, all possessing a vertebral column. Sponges are the only group of animals belonging to the phylum Porifera. A phylum is further divided into classes. Humans belong to the class Mammalia. Living sponges, by contrast, are separated into three different classes. Stepping down the hierarchy, these classes are in turn divided into 25 orders, 127 families and at present around 700 identified living genera. Species from 77 families across the three classes have so far been documented from Australia.

In south-eastern Australia much of the work of identifying sponges was carried out in the 19th century. With only a fraction of those species accurately described in the literature, it is difficult at present to identify the majority of sponge species from south-eastern Australian waters beyond the level of genus.

Characters used for identification

The process of sorting and identifying sponges is based on a set of taxonomic characters; those visible by eye (growth form, colour, texture, surface appearance, oscule location, size and shape), and those by microscope (spicule types and fibre construction). Descriptive terms and definitions for all of these features may be found in *The Thesaurus of Sponge Morphology* (Boury-Esnault & Rutzler, 1997). Complete sets of spicule types for each of the three sponge classes are listed and illustrated in the *Zoological Catalogue of Australia, Volume 12* (Hooper & Wiedenmayer, 1994), and illustrations are reproduced with permission throughout this guide.

External features such as colour and general body shape have limitations as they can vary considerably within a species, depending on factors such as growth history, water current exposure, substrate type and biotic associations (i.e. influence of bacteria or algae).

Molecular technologies are now being implemented to help sort out the continuing difficulties inherent in sponge identification.

Collection and preservation

Sponges have a reputation for being a difficult group to study. A few of the very common sponge species may be identified in the field, but definitive identifications for most species require collection of the sponge or at least a representative fragment. Following collection, the sponge material must be **fixed** either by freezing to retain colour, or by placing immediately into an 80–90 % ethanol solution. Alternatively, sponges may be air-dried for preservation, but in doing so most will lose their colour and many will lose their shape.

A photographic image and a detailed description of the live sponge ***in situ*** (in its natural location), or immediately following collection, are also desirable. The majority of the sponges presented in this guide were photographed in the wild in south-eastern Australia, with samples taken and lodged as voucher specimens at Museum Victoria.

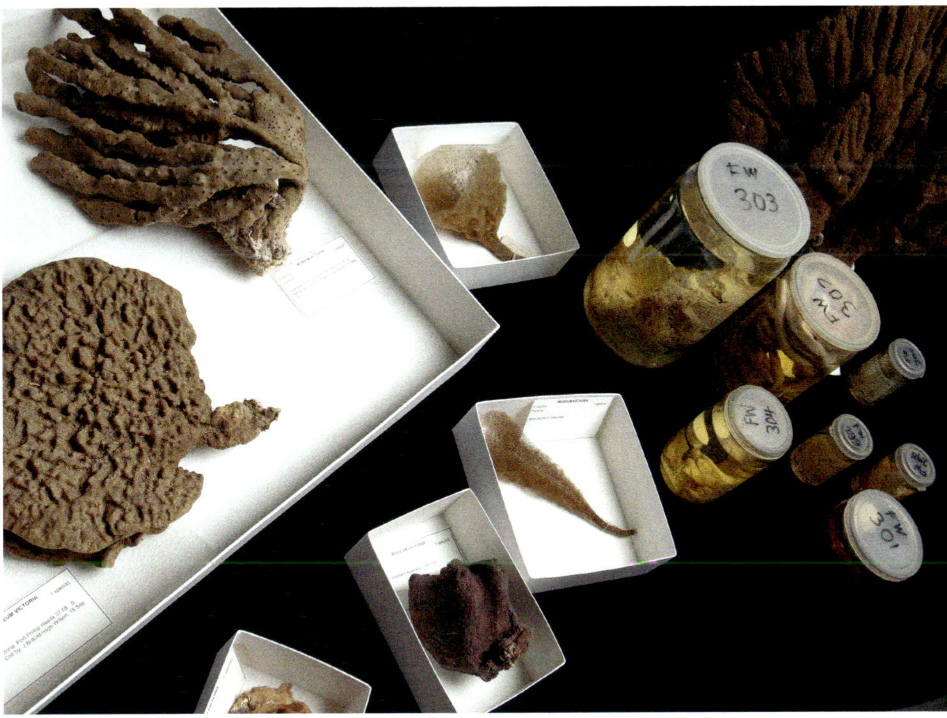

Preserved sponge samples in museum collections enable comparison of vouchered specimens cross-referenced against microscope slides. David Paul

Slide preparation

Most sponge identifications require two types of microscope slide preparations in order to study details of the spicules, if present, and their place in the skeleton. The first is a spicule preparation, necessary to determine the type and shape of the spicules. The second is a cross-section preparation necessary to observe the spicules *in situ*, attributes of the fibre skeleton if present, and any regional differentiation within the body of the sponge.

A step-by-step guide to laboratory preparation of spicules can be found in the Appendix, p. 123.

Keys to sponges

A taxonomic key is a tool used by researchers to identify organisms to a certain level of organisation. A key consists of a series of questions that directs the reader towards the correct combination of characters at each taxonomic level (*i.e.* class, order, family, genus) and ultimately to a species. Formal keys used to identify sponges exist but, due to the difficulties inherent within sponge taxonomy, they must be used with a degree of caution! Sponges are notoriously plastic in their growth form, and examples exist across the phylum of individuals with a tendency either to lose some of the characters on which the keys are based, or to have them masked by the incorporation of sand and detritus into their skeleton, including **foreign** spicules not **native** to the sponge.

Key to the classes of living sponges (Phylum Porifera)

1. Mineral skeleton absent .... Demospongiae
— Mineral skeleton includes a basal skeleton of solid limestone .... (2)
— Mineral skeleton consists of discrete spicules .... (3)
2. Soft parts contain siliceous spicules .... Demospongiae
— Soft parts contain calcareous spicules (test with acid) .... Calcarea
3. Spicules siliceous, larger ones **triaxone** (3 axes) or **hexactine** (6-rayed) .... Hexactinellida
— Spicules siliceous, larger ones **tetraxone** (4 axes) or **monaxone** (single axis) .... Demospongiae
— Spicules calcareous (test with acid), usually **triactine** (3-rayed) or **tetractine** (4-rayed) .... Calcarea

Taxonomic keys beyond the class level are not included here since they are lengthy and based to a large extent on characters requiring microscopic techniques. Such keys may be found in *Systema Porifera – A Guide to the Classification of Sponges* (Hooper & Van Soest, 2002).

About this guide

The sponges of south-eastern Australia are a diverse and largely undocumented group. Only some of the more commonly seen sponges are included here, those for which clear, underwater images are available. Our aim is to extend the range as images and samples are collected in the future.

Each species is introduced with a short description of the key identifying features of the order to which it belongs. This is followed by a description of its family characteristics. Finally, there is a brief summary of the features unique to its genus. Species that are as yet undocumented in the taxonomic literature appear as a genus name followed by a species number, for example *Oscarella* sp. LG1. The numbering of species within a genus is not consecutive since it corresponds to the numbers allocated to sponges in the database of the primary author (a larger collection of sponges than that presented here). It is hoped that these species will ultimately become part of a nationwide database currently being constructed.

The common sponges

Class Demospongiae

Sponges in this class account for approximately 85 percent of all living sponge species. The vast majority of demosponge species are marine and between them they inhabit all of the oceans of the world. The freshwater sponges also belong to this class and likewise are geographically widespread. At present, the demosponges are grouped into 15 orders, 88 families and about 500 genera.

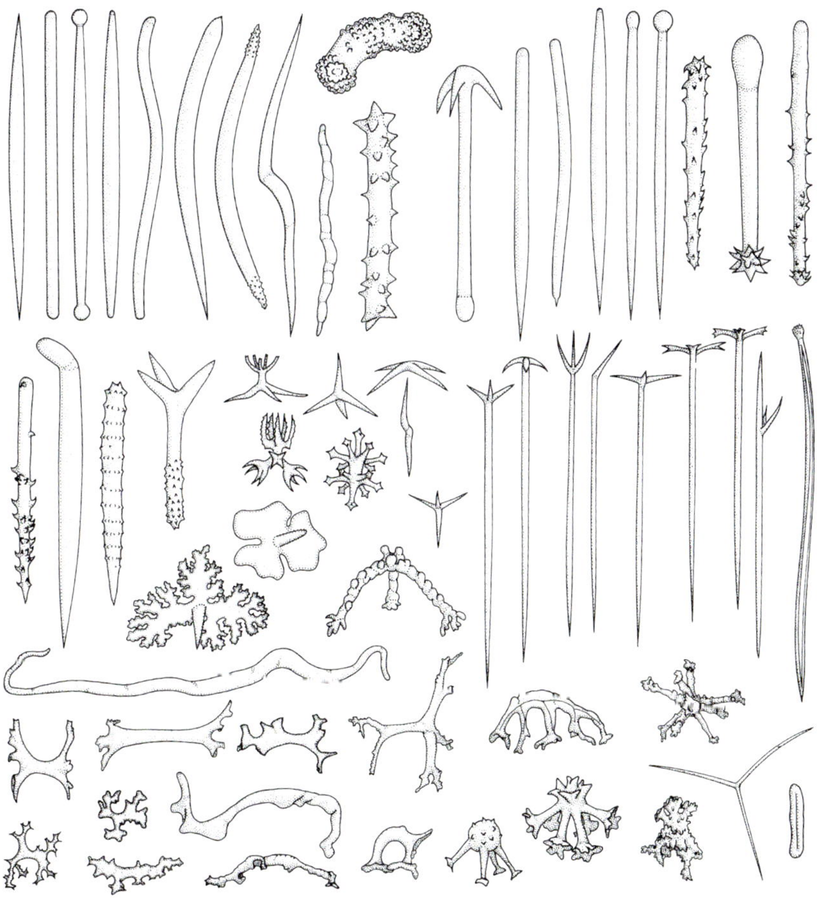

Spicules of class Demospongiae – megascleres

Within this class there are more than 100 different spicule morphologies. The different spicule types and combinations present, and their location within a sponge skeleton, are important characters used to help determine to which genus and species a sponge belongs.

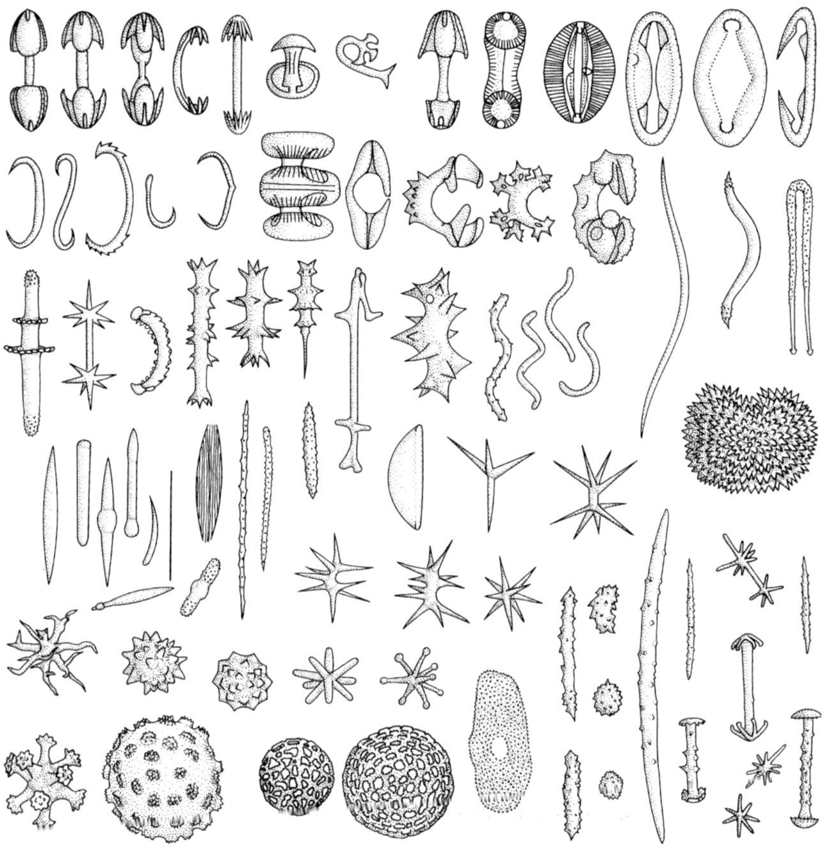

Spicules of class Demospongiae – microscleres

Order Homosclerophorida

Sponges in this group are found mainly in shallow waters. They may be encrusting, rounded, tubular, or occasionally massive. They are usually smooth to the touch, and dense but compressible. Spicules if present are unique to this order and known as calthrops. This order contains only one family, with seven genera, four of which are represented in Australian waters.

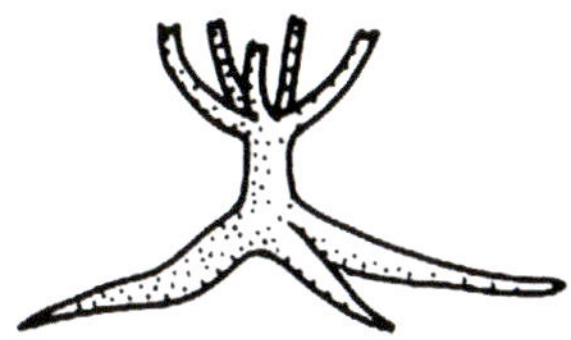

A calthrops megasclere, unique to the order Homosclerophorida.

Family Plakinidae

Members of this family are often encrusting over the substrate. They may be tubular and soft without spicules, or massive and firm, with a skeleton of four-rayed calthrops.

Genus *Oscarella*

Members of this cosmopolitan genus are characterised by the absence of a skeleton of any form (either **spicular** or fibrous) and by the low ratio of volume of collagenous tissue to volume of chambers (approximately 1:1). The tubular species shown here encrusts over boulders.

Oscarella sp. LG1 (Julian Finn), with inset picture of external tubular canals (Mark Norman), Portsea Pier, Port Phillip Bay.

Order Astrophorida

Sponges of the five families in this order are characterised by their **asterose microscleres** (small star-shaped spicules), often in combination with tetractinal megascleres (large four-rayed spicules), that are peculiar to this order. The megascleres tend to be arranged radially, obvious at least in the **cortex** (peripheral region) of the sponge.

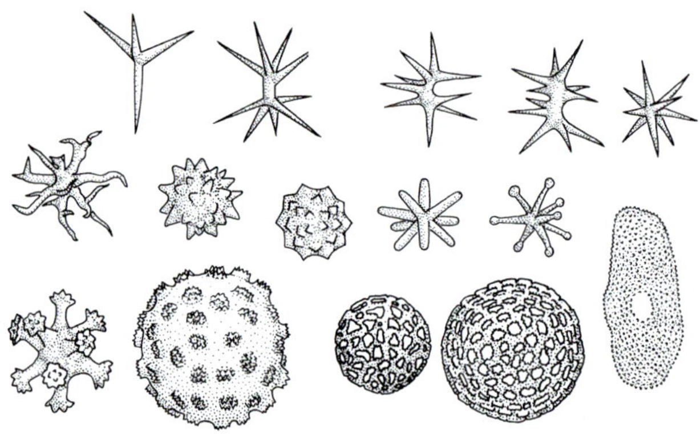

Asterose microscleres

Family Ancorinidae

The defining characters for this family and its 15 genera are the presence, or absence, of certain types of spicules. They have **triaene** megascleres (large pitchfork-like spicules with a long shaft, one end pointed and the other end triple-pronged), and one type or a combination of microscleres including asterose forms "and/or **microrhabds** (spined rods)". Ancorinid sponges adopt diverse growth forms including encrusting, massive and spherical, some with long, inhalant and/or exhalant tubes.

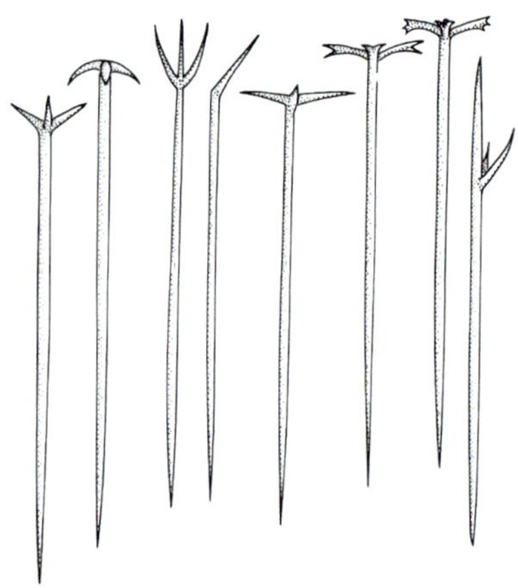

Triaene megascleres

Genus *Stelletta*

Sponges of this cosmopolitan genus are generally massive to spherical in shape with a cortex rich in collagen. Representatives of the genus occur around the southern Australian coast at depths of 10–160 metres.

Stelletta tuberculata, Symonds Channel sponge garden, Port Phillip Bay. Mark Norman

Genus *Psammastra*

The surface of these sponges is typically covered with small bumps or cone-shaped projections that are raised up from the skeleton beneath, a feature known as **conulose**. The sponges are often covered with a fine layer of adhering sand grains. Representatives of the genus occur throughout the Pacific Ocean and in Port Phillip Bay, in sandy or muddy habitats from depths of 2 metres.

Psammastra sp. LG1, Symonds Channel sponge garden, Port Phillip Bay. Mark Norman

Genus Ecionemia

Sponges of this genus are either encrusting in thick mats or massive to spherical in shape but, unlike most other genera in this family, they do not have a distinct cortex. Representatives of this quite common genus occur around almost the entire Australian coast, with only one species currently recorded from Victorian waters.

Ecionemia robusta showing large exhalant oscule, Nepean Wall, Port Phillip Bay. Mark Norman

Order Hadromerida

Hadromerid sponges are often brightly coloured and are also characterised by a radial arrangement of megascleres in the cortex. The megascleres are pin-shaped **tylostyles** and pointed, rod-shaped **oxeas**. Microscleres if present may be star-shaped, rod-shaped or spiral and are key characters when separating these sponges at the family level.

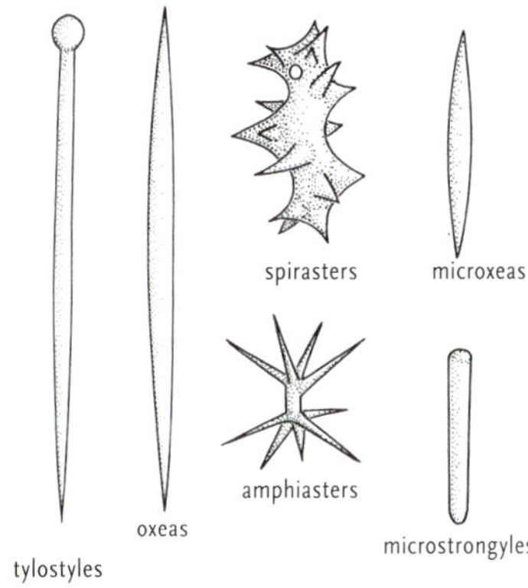

Spicule types found in sponges of the Order Hadromerida.

Family Clionidae

Sponges in this family are able to excavate into limestone or calcareous substrates. They possess tylostyles, megascleres and a variety of microscleres.

Genus *Cliona*

Primarily excavating, these sponges do occasionally appear above the substrate as an irregular lump or massive form, like the one shown here. Megascleres take the form of tylostyles, while the microscleres are **spirasters** that may be straight, bent, spiralled or spiny.

Cliona sp. LG1, Wilsons Promontory. Mark Norman

Genus *Spheciospongia*

These sponges are massive in growth form with specialised incurrent pore sieve plates. Due to the loss of red light with depth, these sponges appear a deep royal blue in colour, their purple colour only being visible with camera lights or in sunlight. This species occurs quite commonly in Victorian coastal waters and across the Great Australian Bight to Western Australia.

Spheciospongia purpurea with close-up inset, Lonsdale Wall, Port Phillip Bay. Mark Norman

Spheciospongia sp. LG1 (Mark Norman), with close-up inset (Julian Finn), Nepean Wall, Port Phillip Bay.

Spheciospongia papillosa (Mark Norman), with inset showing sieve plates (Julian Finn), Nepean Wall, Port Phillip Bay.

Family Spirastrellidae

Spirastrellids are primarily encrusting sponges, capable of excavating limestone. They possess a layer of large spiraster microscleres in the cortex. The species shown here appears to be closely associated with an encrusting octocoral, giving it a two-tone appearance.

Spirastrella sp. LG1, with encrusting octocoral, Whaleback Reef, Point Hicks. Mark Norman

Family Suberitidae

Suberitid sponges are characterised by their lack of a differentiated cortex, and by the absence of microscleres in all but a few species. Body form can be encrusting, massive, globular, or stalked and branching. Of the 11 genera in this family, two are included here.

Genus *Protosuberites*

Members of this genus all have an encrusting form and a velvety surface, barely visible by eye, produced by a brush of protruding tips of tylostyle spicules at the sponge surface. This characteristic velvety appearance is termed **hispid**, and appears repeatedly throughout the phylum Porifera. The sponge pictured takes the form of a thin layer over a large boulder.

Protosuberites sp. LG1, Mornington Pier, Port Phillip Bay. Mark Norman

Genus *Suberites*

Three very different growth forms within this genus are depicted here, but all species of *Suberites* have the characteristic hispid surface. This genus is cosmopolitan but most species are found in cooler temperate waters like those off southern Australia's coast. The pale yellow to cream-coloured *Suberites globosus* shown here is commonly seen attached to rocky substrate amongst seagrass or algae and has been recorded from Victoria, Tasmania and New South Wales.

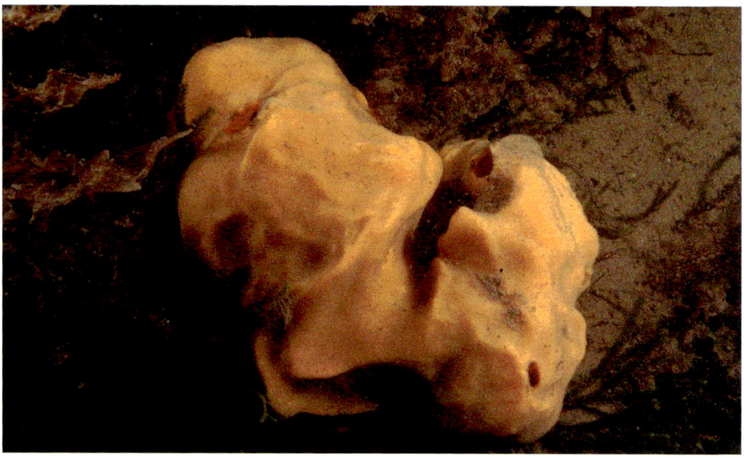

Suberites globosus, Mornington Pier, Port Phillip Bay. Mark Norman

Suberites sp. LG1, Mornington Pier, Port Phillip Bay. Mark Norman

Suberites sp. LG2 with close-up inset, Whaleback Reef, Point Hicks. Mark Norman

Family Tethyidae

Sponges in this family have **styles** (rod-shaped megascleres, rounded at one end and pointed at the other) and two different-sized asterose microscleres. This family is highly diverse in tropical latitudes and occurs at depths from 0 to 2000 metres, although most known species occur at less than 50 metres. Of the 14 genera in the family only one, *Tethya*, is found in Victorian waters.

Genus *Tethya*

These distinctive, often spherical sponges are commonly called 'golf ball' sponges. They are common in southern Australian waters and are orange-red, yellow or pale pink. Some have root-like anchoring structures known as **pedicels** (see image of *Tethya* sp. LG2) for attachment to the substrate. Members of this genus have a firm, almost incompressible texture due to the dense, radial arrangement of spicules and a warty, **tuberculose**, surface. Almost a third of the world's known species of *Tethya* occur in Australia, with 15 species found in southern Australia. Separation of the species of *Tethya* is based on the overall size and shape of the sponge, and the type and placement of the asterose microscleres.

Sponges in the genus *Tethya* often exhibit budding propagules, a form of asexual reproduction whereby the parent sponge produces a stalk of spicules at its surface, terminating in a bud that detaches and floats away to become a separate individual, (see image of *Tethya* sp. LG3).

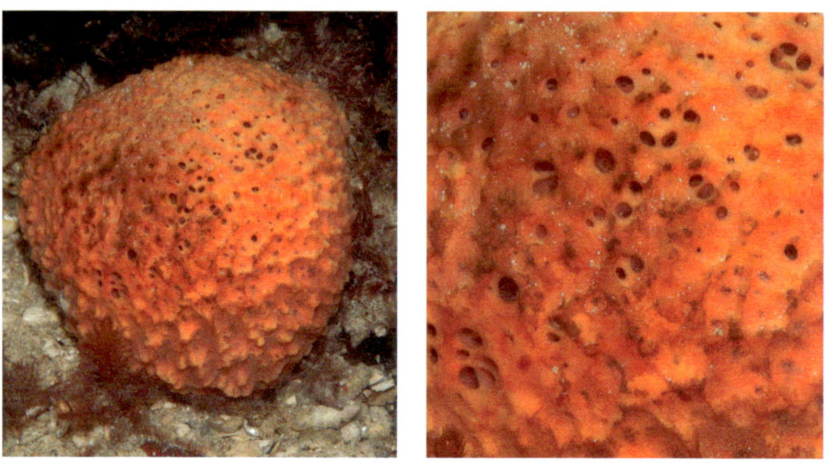

Tethya sp. LG1 with close-up, right, Portsea Pier, Port Phillip Bay. Mark Norman

Tethya sp. LG2 showing pedicels, Mornington Pier, Port Phillip Bay. Mark Norman

Tethya sp. LG3 showing budding propagules, Pillar Point, Wilsons Promontory. Julian Finn

Chondrosia sp. LG1. Mark Norman

Order Chondrosida

Family Chondrillidae

Sponges in this order all belong to the single family Chondrillidae and may be encrusting to massive in growth form. They are widely distributed and tend to be found in shallow water, often in caves or beneath rocks. They have a smooth appearance and a rubbery texture due to their densely collagenous matrix. They have a distinct outer cortex strengthened by **fibrils** of collagen and a complete absence of megascleres. If present at all, spicules are asterose forms of microscleres, and these appear in only one of the four genera in the family, *Chondrilla*.

Genus *Chondrosia*

Sponges in this genus are widespread from tropical to temperate waters and occur mostly at depths of less than 50 metres. They are entirely collagenous, contain no spicules and are most often rounded, **elongate** (drawn out) or **lobate** (having lobes) in shape. The species shown here covers rock surfaces at a depth of 25 metres at Cape Howe in eastern Victoria.

Chondrosia sp. LG1 with close-up, right, Cape Howe Marine National Park. Mark Norman

Order Poecilosclerida

This order contains the highest diversity of sponge species in the phylum, spread amongst 25 families. Member species have been recorded from all marine habitats. Poecilosclerid sponges are characterised by the presence of both organic (fibre) and inorganic (spicular) skeletal components, both usually well developed. Skeletal elements are often localised to particular regions of the sponge. A diverse range of spicule types is displayed throughout this order including the c-shaped **chelae** microscleres that are peculiar to this order. Examples from 11 of the 25 families are included here.

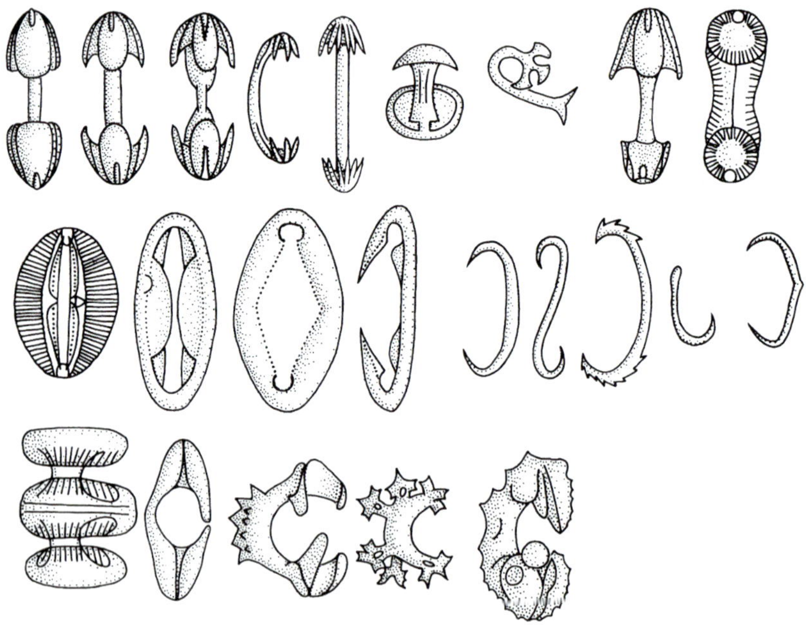

Chelae microscleres

Family Acarnidae

Representatives of this family are distributed worldwide, and many take burrowing, excavating or encrusting forms with protruding fistules, as seen in the sand-burrowing species shown here.

Genus *Acheliderma*

These fistulose sponges are characterised by the possession of elongate microxea microscleres with diamond-shaped tips. The species shown here, *Acheliderma fistulatum*, known only from Victoria and Bass Strait, consolidates sand grains into a solid mass with very fine, transparent fistules protruding.

Acheliderma cf. *fistulatum* showing transparent fistules, Symonds Channel sponge gardens, Port Phillip Bay. Mark Norman

Family Microcionidae

Approximately 460 species of sponges from this family have been described worldwide with about 150 known from Australia, most of which occur in shallow water. Taxonomic identifications within this family are based on microscopic features including regional differences within the skeleton of the sponge and microsclere morphology. Members of this family adopt encrusting, lobate, branching or **lamellate** (plate-like) **habits**.

Genus *Clathria*

Sponges of the genus *Clathria* contain one or two categories of styles present at the surface as a continuous brush or **palisade**. The skeleton is a well-structured reticulation of fibres cored by styles and **echinated** (studded) by **acanthostyles** (spined styles) – the head of the spicule is implanted in the fibre and the pointed end protrudes from it. A great range of growth forms are exhibited in this genus with species spread over seven subgenera, three of which are represented here.

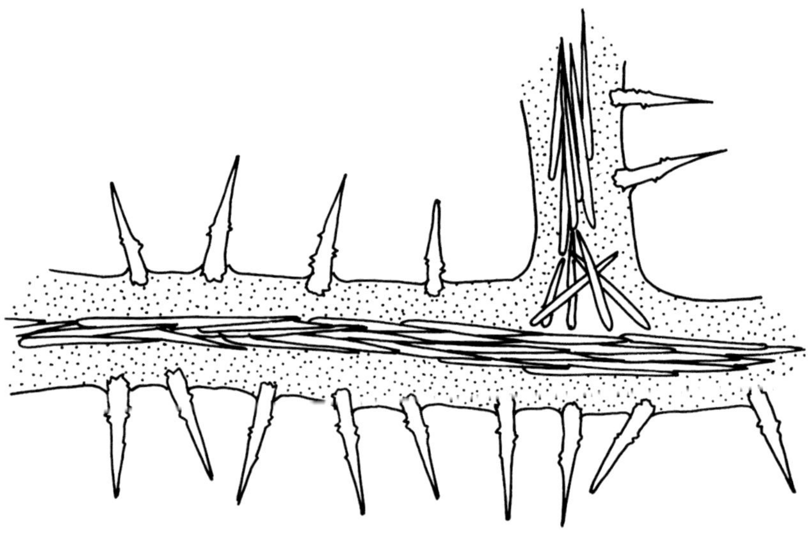

Acanthostyles echinating a sponge fibre.

Subgenus *Clathria (Clathria)*

These sponges may be massive or plant-like in habit. The skeleton is **undifferentiated** (uniform in appearance) throughout the internal region of these sponges. Fibres are usually echinated by megascleres, which may be spined or smooth.

Clathria (*Clathria*) sp. LG7 (Mark Norman) with close-up inset (Julian Finn), Lonsdale Wall, Port Phillip Bay.

Subgenus *Clathria (Isociella)*

The skeleton of sponges in this subgenus is also a uniform mesh of fibre and/or spicular elements, a structure known as **reticulate**. Fibres are *not* echinated in this subgenus. In habit they may be massive, lamellate as seen here, lobate or **digitate** (finger-like).

Clathria (Isociella) sp. LG2, Cape Wellington, Wilsons Promontory. Mark Norman

Subgenus *Clathria (Thalysias)*

These members of the genus *Clathria* have two size-classes of styles at the surface and exhibit echinated fibres in the skeleton. They may be thickly encrusting over the substrate (as seen here) or, as in the other subgenera, massive or **arborescent** in form.

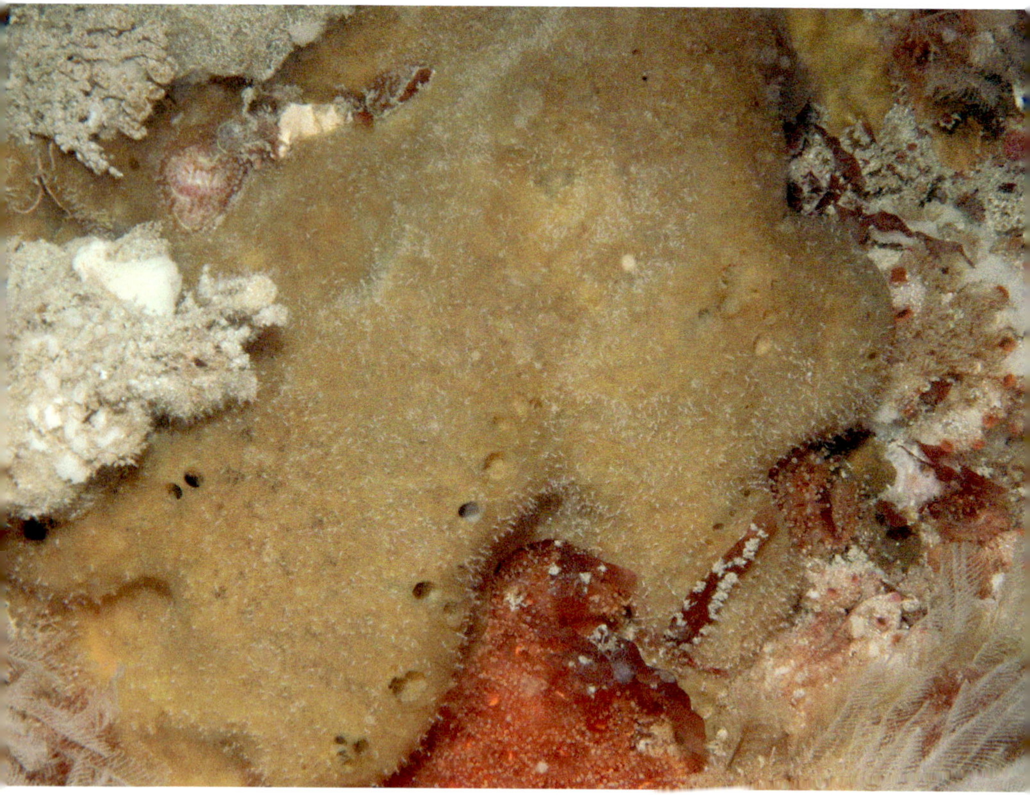

Clathria (Thalysias) sp. LG4, Pillar Point, Wilsons Promontory. Julian Finn

Genus *Echinochalina*

Sponges in this genus display all growth forms including encrusting, branching, massive, and even honeycombed. Two types of megascleres are found in these sponges; one in the outer **ectosome** region, being the same as those found **coring** the fibres of the inner **choanosome**, the other echinating the fibres. Within this genus there are two subgenera, one of which is included here.

Subgenus *Echinochalina (Echinochalina)*

All growth forms are found in this subgenus, including the peculiar honeycombed appearance of the sponge shown here. All megascleres found in these sponges have one pointed end while the other is rounded, a spicule morphology known as **stylote**. If present, microscleres found in this subgenus are claw-like chelae, or small, curved, wave-like **toxas**.

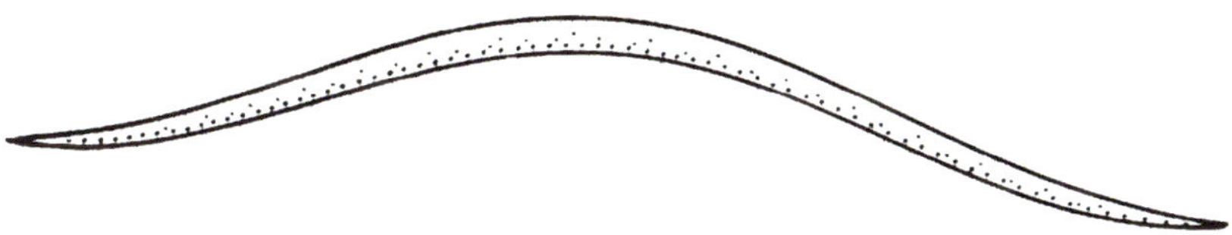

Illustration of a toxa, wave-like microscleres found in only a couple of families of sponges.

Echinochalina (Echinochalina) sp. LG1, Pillar Point, Wilsons Promontory. Julian Finn

Genus *Holopsamma*

Sponges in this genus have a typically reticulate, honeycomb growth form with fibres both cored and echinated by the *same* spicule type, although the coring spicules may be partially or wholly replaced by sand particles.

Of the five species reported from southern Australia, only one is included here. *Holopsamma laminaefavosa* is commonly found around the Victorian coast, Tasmania and New South Wales from depths of 1–100 metres. It tends to adopt a thickly encrusting to lobate growth form and varies in colour from orange-brown to cream.

Holopsamma laminaefavosa, Pillar Point, Wilsons Promontory. Julian Finn

Genus *Echinoclathria*

These sponges may be encrusting, lobate, bushy, digitate, or **flabelliform** (fan-shaped) in growth form, as seen here. Like the other microcionid sponges, the skeleton is a mesh of fibres, here cored and echinated by the same type and size of stylote megascleres. A radial skeleton is also visible microscopically at the periphery, where the spicules appear to fan out towards the edge of the sponge.

Echinoclathria sp. LG1, Cape Wellington, Wilsons Promontory. Mark Norman

Family Raspailiidae

This family exhibits a wide range of growth forms and is typically represented by sponges with a pronounced hispid surface, produced by the tips of spicules, usually styles or oxeas, protruding through the sponge surface. These large protruding spicules, sometimes visible by eye, are typically surrounded by bouquets of smaller spicules at their base – a feature termed **raspailiid**. The fibre skeleton is usually reticulate and may be compressed in the peripheral region.

Raspailia sp. LG1, Portsea Pier, Port Phillip Bay.
Julian Finn

Genus *Raspailia*

This cosmopolitan genus contains member species found from the tropics to Antarctic waters over a wide depth range. These sponges exhibit growth forms including arborescent, massive, and thickly encrusting to lobate, like the sponge seen here growing over pier piles at Portsea.

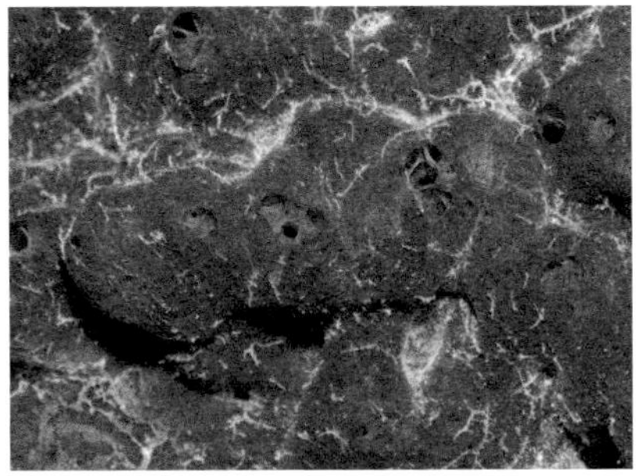

Raspailia sp. LG1, close-up of sponge on left.
Mark Norman

Family Chondropsidae

Chondropsid sponges may be encrusting, massive, flabellate or digitate in form and are characterised by their **vestigial** (much reduced) megascleres, which are usually **strongyles** (rod-shaped, rounded at both ends) or occasionally styles (pointed one end and rounded the other). If present, the microscleres found in this family include chelae and/or c- and s-shaped **sigmas**. These sponges often incorporate large amounts of sand and broken spicule fragments into their skeletons, drawn in with the incurrent of seawater. There are five genera in this family, all of which have been reported from Victorian waters. Examples of two of the more commonly occurring genera are included here.

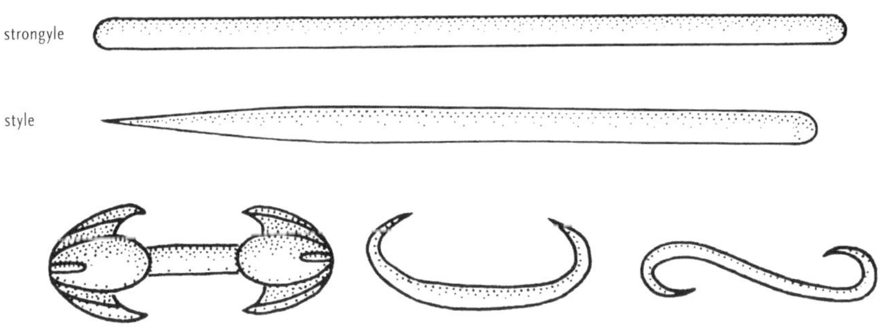

Spicules found in sponges of the family Chondropsidae.

Genus *Chondropsis*

Sponges in this genus are characterised by their tendency to incorporate sand and foreign spicule fragments into the skeletal fibres. Native spicules, if seen in these sponges, are vestigial strongyles.

Chondropsis sp. LG1, Symonds Channel sponge gardens, Port Phillip Bay. Julian Finn

Chondropsis cf. *kirki* with close-up inset, Symonds Channel sponge gardens, Port Phillip Bay. Julian Finn

Chondropsis cf. *kirki*, Portsea Pier, Port Phillip Bay. Julian Finn

Chondropsis sp. LG2, Pillar Point, Wilsons Promontory. Julian Finn

Chondropsis sp. LG3, Pillar Point, Wilsons Promontory. Julian Finn

Chondropsis sp. LG4, Nepean Bay, Port Phillip Bay. Mark Norman

Chondropsis sp. LG5, Nepean Bay, Port Phillip Bay. Mark Norman

Chondropsis cf. *kirki*, Point Hicks, Whaleback Reef. Mark Norman

Chondropsis sp. LG7, Point Hicks, Whaleback Reef. Mark Norman

Genus *Phoriospongia*

Sponges in this genus are also characterised by the incorporation of sand and foreign spicule fragments into their skeletal fibres. The main difference between sponges in this genus and those of the genus *Chondropsis* are the megascleres. While vestigial, the native spicules secreted by the sclerocyte cells of *Phoriospongia* are styles, not *strongyles*.

Phoriospongia sp. LG2, Symonds Channel sponge gardens, Port Phillip Bay. Mark Norman

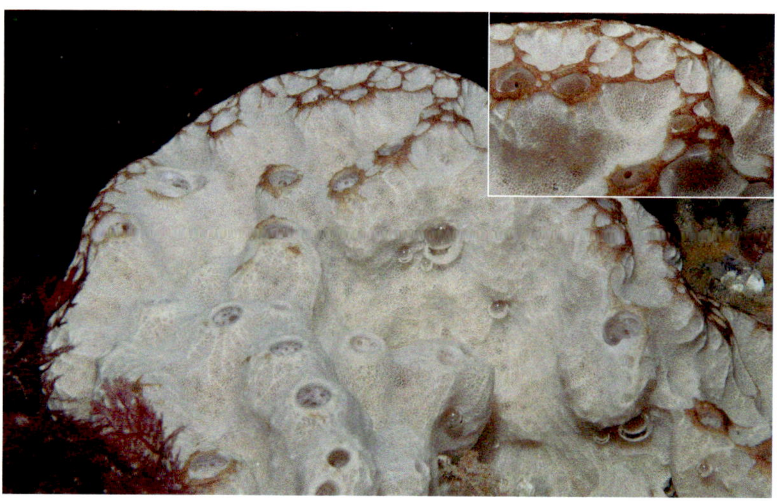

Phoriospongia sp. LG3 with close-up inset, Cape Wellington, Wilsons Promontory. Julian Finn

Genus *Psammoclema*

These chondropsid sponges contain no native spicules at all – instead the skeleton consists only of sand and foreign spicule debris. They tend to be either massive in form or spreading over the substrate.

Psammoclema sp. LG2, Nepean Bay, Port Phillip Bay. Mark Norman

Family Coelosphaeridae

Members of this family adopt various growth forms from massive to encrusting, branching to **fistulose** (having hollow tubes). The seven genera in this family are grouped together because of the similarities in their mesh-like skeletal structure and the combinations of spicules, which include **tylotes** (rod-shaped megascleres with slightly swollen ends) in the outer layer and chelae (claw-like microscleres).

Genus *Lissodendoryx*

This cosmopolitan sponge genus contains many species and may be massive, lobate or flabellate, typically with an irregular surface. The species shown is thickly encrusting to conical, here seen growing on pier piles.

Lissodendoryx sp. LG1, Mornington Pier, Port Phillip Bay. Mark Norman

Family Crellidae

Members of this family adopt many habits ranging from encrusting to club-shaped, branching to massive. The group is characterised by an outer crust of spined spicules, either oxeas or styles, lying **tangential** (parallel) to the surface. Smooth structural spicules form a regular mesh deeper in the sponge. Most crellid species exhibit specialised **areolae** (pore sieve-plates, or groups of inhalant pores) on the surface. Identification of genera and species within this family relies on megasclere morphologies throughout the sponge and the presence or absence of certain microscleres.

Genus *Crella*

Sponges in this genus may be lobate, stalked or encrusting and are found throughout the world from shallow coral reefs, such as the Great Barrier Reef, to the deep waters of Antarctica. Encrusting species often settle on the shells of bivalve mollusc such as the one shown here, the Doughboy Scallop (*Mimachlamys asperrima*), or are cut off by decorator crabs such as the Great Spider Crab (*Leptomithrax gaimardii*) and worn on the crab's carapace. In both cases the host animals gain camouflage and chemical protection, while the sponge obtains mobility and may gain better access to food-rich waters (see ***Biotic associations***).

Crella (Pytheus) cf. *incrustans* as defensive coat of the Doughboy Scallop (with blue eyes), Portsea Pier, Port Phillip Bay. Mark Norman

Family Hymedesmiidae

Like members of the family Crellidae, most species in the family Hymedesmiidae feature specialised areolae on the surface. They differ from the crellids by lacking a crust of spined spicules near the surface. Again, these sponges may be encrusting, branching or massive.

Genus *Phorbas*

These sponges have a distinct crust of chelae microscleres at the surface. The bright orange species depicted here can be thickly encrusting or lobate in form, and is commonly found growing on pier piles.

Phorbas sp. LG1 (Julian Finn) with close-up showing encrusting lace coral (Mark Norman), Portsea Pier, Port Phillip Bay.

Genus *Pseudohalichondria*

These massive or flabellate sponges are made up of a network of thick fibres cored by **subtylostyles** (smooth megascleres with one end pointed and a subterminal swelling at the other) and sand particles incorporated into their skeletal framework. Microscleres in this genus include the unusual, spined chelae.

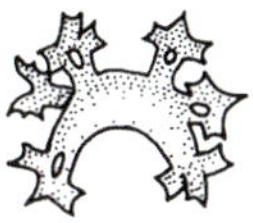

Subtylote megasclere and spined chelate microsclere.

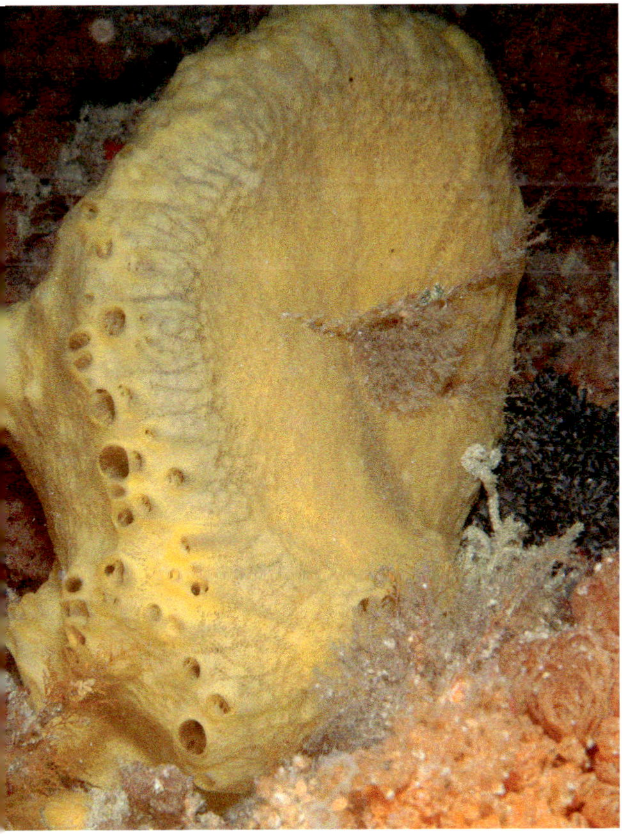

Pseudohalichondria sp. LG1, Pillar Point, Wilsons Promontory. Julian Finn

Family Iotrochotidae

Members of this family adopt the full range of growth forms but are distinguished by their possession of unique double umbrella-shaped microscleres known as **birotules**. A range of megascleres are also found in this family, with the exception of the genus presented here, in which there are none.

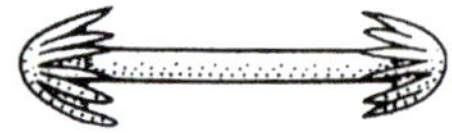

Umbrella-shaped birotule microsclere

Genus *Iotrochopsamma*

While the unique birotule microscleres are present in sponges of this genus, larger structural megascleres are completely lacking. These sponges incorporate sand grains into their skeletons for strength. Often branching in form, they have a porous, sandy surface appearance.

Iotrochopsamma cf. *arbuscula*, Pillar Point, Wilsons Promontory. Julian Finn

Family Tedaniidae

These sponges are characterised by small, spined, rod-shaped microscleres known as **onychaetes**. The chelae, typical of poecilosclerid sponges, are not found in this family. Growth forms range from encrusting to massive or digitate. Structural spicules of these sponges may be enclosed within spongin fibres in the skeleton or, where fibres are lacking, may be cemented together at their tips.

Onychaete microscleres found exclusively in sponges of the family Tedaniidae.

Genus *Tedania*

Members of this genus are characterised by having different forms of megasclere spicules in the outer ectosome rather than deeper in the choanosome. The species shown here, *Tedania (Tedania) anhelans*, is commonly bright orange and can be thickly encrusting to cushion-shaped with conspicuous oscules clustered on top of the lobes. It has been recorded from coastal waters around Australia.

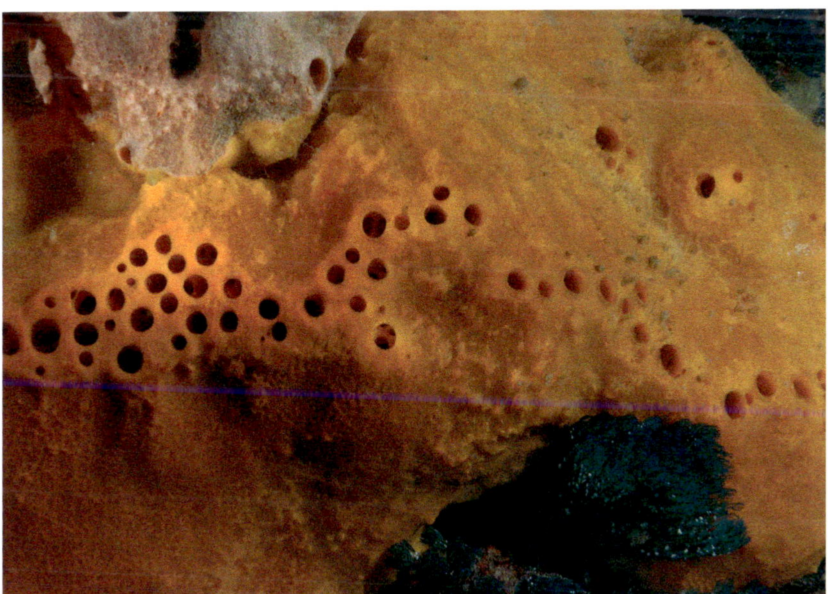

Tedania (Tedania) anhelans, Portsea Pier, Port Phillip Bay. Julian Finn

Tedania (Tedania) anhelans, Port Phillip Bay. Julian Finn

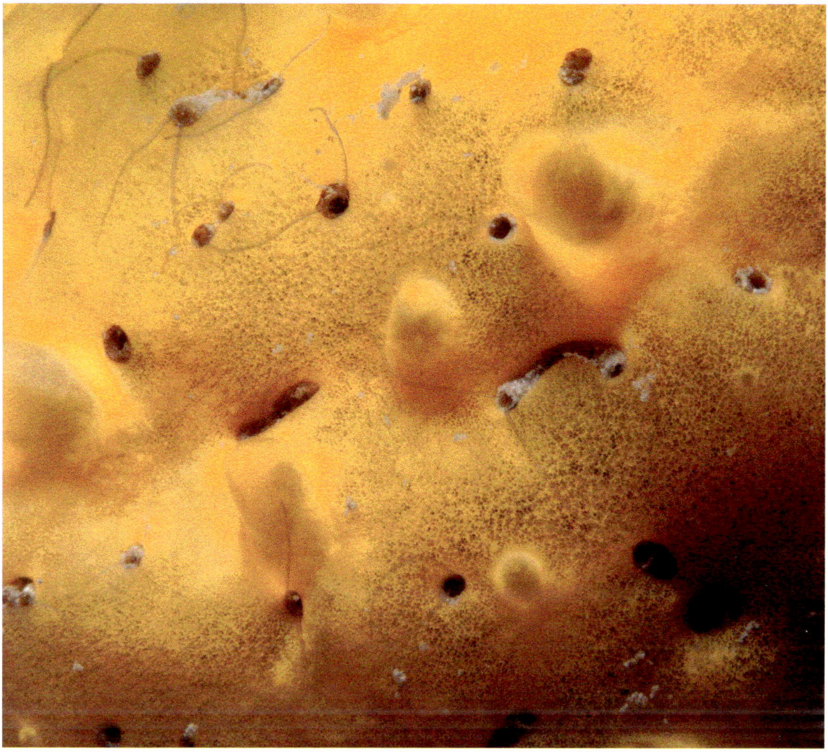

Tedania (Tedania) sp. LG1, close-up showing worm tubes and antennae. Julian Finn

Family Desmacellidae

Sponges in this family lack the chelae microscleres that are characteristic of most other families in the Order Poecilosclerida. Instead they have a diverse range of microscleres that may include sigmas, comma-like **commata** and fine, hair-like **raphides**. This family contains six genera that exhibit examples of the full variety of growth forms from encrusting to **caliculate** (cup-shaped), branching to massive. Some species in this family are known to cause severe dermatitis reactions of the skin on contact.

Genus *Biemna*

There are over 50 species in this genus occurring in oceans worldwide and over a large depth range. Only a few species are reported to occur off the southern Australian coast. The pale yellow species shown is cushion-shaped and was found on pier piles.

Biemna sp. LG1 showing associated worm tubes, Mornington Pier, Port Phillip Bay. Mark Norman

Family Mycalidae

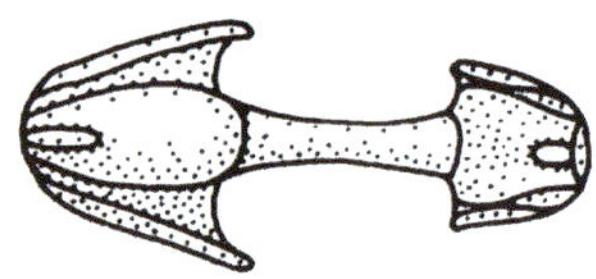

Asymmetrical anisochelae, characteristic of sponges of the family Mycalidae.

Sponges in this family may be encrusting, massive, branching or flabellate, but the species treated here are restricted to those having a combination of two features; a surface skeleton of spicules lying tangential to the surface of the sponge, and possession of asymmetrical chelate microscleres, known as **anisochelae**. Superficial sculpturing of these sponges is often visible through the transparent surface in the form of grooves and ridges.

Genus *Mycale*

Mycalostyles, megascleres exclusive to sponges of the family Mycalidae.

The larger of the two genera in this family, *Mycale,* is further sub-divided into 11 subgenera and about 200 species. Sponges of this genus are characterised by their **mycalostyles**, unique rod-shaped megascleres that have a slight constriction at the neck of the rounded end. These sponges are usually soft in texture but have a high fibrous content. Examples of two subgenera are included here.

Subgenus *Mycale (Arenochalina)*

Species of the subgenus *Mycale (Arenochalina)* characteristically incorporate sand and foreign material and/or filamentous algae into their fibre skeletons. The pale orange species (sp. LG1) was found encrusting pier piles and is itself encrusted by filter-feeding worms.

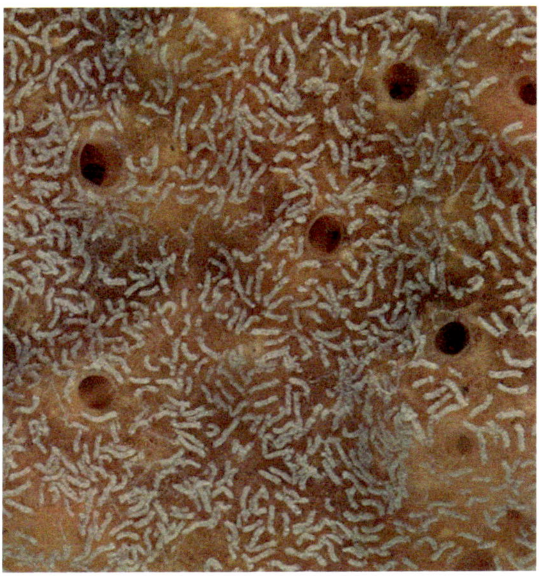

Top, *Mycale (Arenochalina)* sp. LG1, Portsea Pier, Port Phillip Bay. Below, close-up of same sponge showing associated worms. Julian Finn

By contrast, the second member of this subgenus treated here (sp. LG2) is a stalked, branching sponge, found attached to hard substrate in Symonds Channel in southern Port Phillip Bay.

Mycale (Arenochalina) sp. LG2, Symonds Channel sponge gardens, Port Phillip Bay. Mark Norman

Subgenus *Mycale (Grapelia)*

Members of this subgenus are distinguished by **unguiferous** anisochelae, asymmetrical chelae microscleres that have tooth-like projections at each end. These microscleres are often arranged in a rosette. The species shown here, *Mycale (Grapelia) australis*, is massive-spherical in growth form with a warty surface and may also have **conules** (small conical protuberances). It occurs across southern Australia from New South Wales to Western Australia, at depths of 4–40 metres.

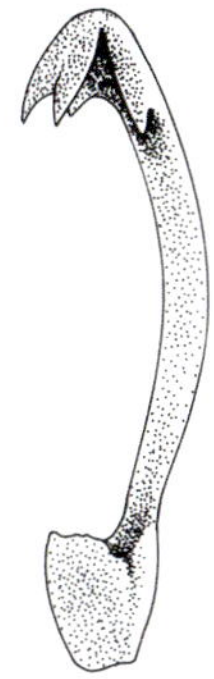

Tooth-like projections of the unguiferous anisochelae.

Mycale (Grapelia) australis, Symonds Channel sponge gardens, Port Phillip Bay. Mark Norman

Order Halichondrida

Sponges in this order exhibit the full range of growth forms and textures. They can possess any combination of three megasclere types (styles, oxeas and strongyles) without showing any regionalisation within the skeleton. The megascleres in members of this order are often **sinuous** (curved or snake-like), or deformed in some way. Microscleres are generally absent but if present tend to be **microxeas** or hair-like **raphides**. Two of the five families are represented here.

Family Axinellidae

This family contains around 300 species worldwide and examples are found from the shallowest depths down to at least 1800 metres. Axinellid sponges typically have a velvety surface, caused by the protruding points of surface spicules. They are commonly red, orange or yellow and may be encrusting, massive, branching or flabellate (fan-shaped).

Genus *Dragmacidon*

Sponges of this genus are unbranched and may be club-shaped, shrub-like or thickly encrusting. The surface is smooth, not velvety, but is covered in short, thick processes or **tubercles**.

Dragmacidon cf. *clathriforme*, Cape Howe Marine National Park. Mark Norman

Genus *Phakellia*

These sponges are either lamellate or flabellate and are usually stalked. The surface is velvety and often imprinted with 'veins' formed by the main tracts of spicules beneath the surface. The skeletal structure is provided by tracts (columns) of sinuous strongyles woven together to form an axis, with a connecting network formed by lines of single spicules. This sponge was found loose on the sand, probably recently broken loose from its point of attachment.

Phakellia sp. LG1, Flinders, Western Port Bay. Mark Norman

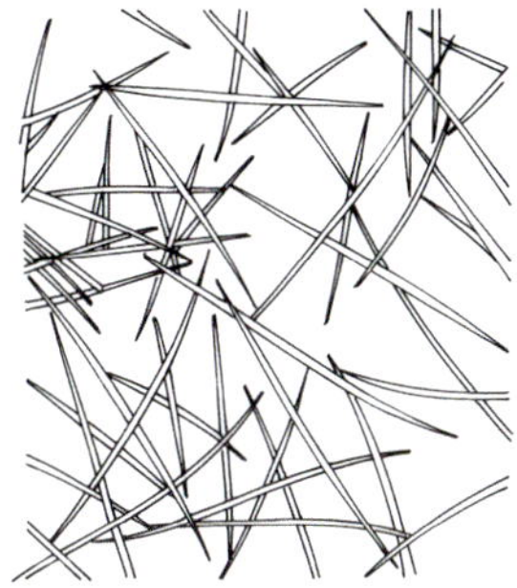

Confused spicule arrangement known as halichondroid.

Family Halichondriidae

Sponges in this family are distributed widely throughout the world's oceans, predominantly in shallow, coastal waters. They are distinguished by their apparent *lack* of organisation of the skeleton. The megascleres (styles, oxeas or both) are distributed in a random, criss-crossed arrangement throughout the sponge. The descriptive term **halichondroid** is used to describe this apparent haphazard arrangement of spicules, even when found in sponges of other groups. Spongin fibres are rare and collagenous material is often reduced. Spicule density is high, resulting in sponges with a firm, harsh or brittle consistency. Microscleres are absent in this family of sponges, with the exception of raphides which are found in only one of the 11 genera.

Genus *Ciocalypta*

These sponges typically have a massive base buried beneath a soft substrate, from which protrude semi-transparent, tapering fistules. The ectosome is easily detachable as a skin-like layer. Members of the genus *Ciocalypta* are commonly found in sandy or muddy areas like Port Phillip Bay.

Left, *Ciocalypta* sp. LG1, Symonds Channel sponge gardens, Port Phillip Bay. Right, close-up of same sponge. Mark Norman

Ciocalypta cf. *massalis*, Cape Wellington, Wilsons Promontory. Mark Norman

Order Haplosclerida

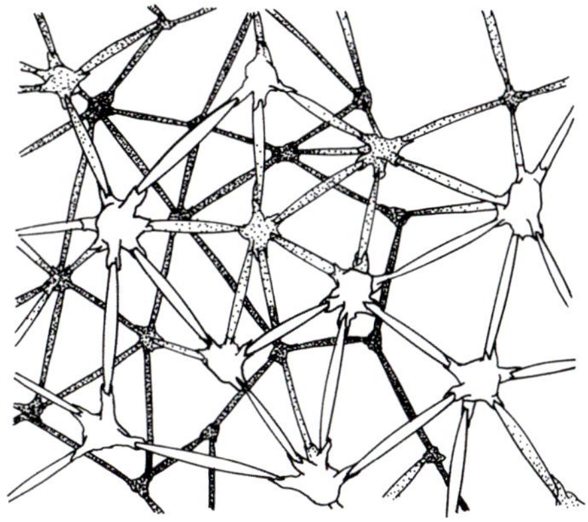

Isodictyal skeletal arrangement

This order contains the greatest biodiversity of sponges with representatives inhabiting all oceans and habitat types, including freshwater bodies, of every continent in the world. Sponges of this order are characterised by the triangular mesh formed by their fibres and/or spicular skeletal elements. This is termed an **isodictyal** skeleton. The megasclere spicules found in haplosclerid sponges are exclusively oxeas or strongyles. Microscleres are present in representatives of only a few genera. Some of the more common marine species and one freshwater sponge are included here.

Family Callyspongiidae

Most growth forms are represented in this family of sponges including encrusting, massive, vase-shaped, tubular, flabellate, lamellate and branching. Species within the family are differentiated by two main features: the reticulation of spicules (mesh) in the ectosome, and the presence or absence of spongin fibres in the choanosome.

Genus *Callyspongia*

This large genus is further separated into five subgenera, two of which are included here. All of these sponges are characterised by an abundance of spongin fibres in their skeleton, making them both tough and flexible. The mesh or reticulation is formed by a **hierarchy** of fibres; **primary**, **secondary** and **tertiary**. These sponges are generally pale in colour and the skeletal fibres are clearly visible through the transparent surface. Some of the many growth forms found in this genus are depicted here.

Subgenus *Callyspongia (Callyspongia)*

Members of this subgenus are characterised by having a uniform mesh size in the outer ectosome. The fibres within the deeper choanosome form a rectangular mesh and are cored by tracts of spicules. There are no free spicules seen in the microscopic sections of these sponges.

Callyspongia (Callyspongia) sp. LG1, Symonds Channel sponge gardens, Port Phillip Bay. Julian Finn

Callyspongia (Callyspongia) sp. LG2, Pillar Point, Wilsons Promontory. Julian Finn

Callyspongia (Callyspongia) sp. LG3, Nepean Bay, Port Phillip Bay. Mark Norman

Callyspongia (Callyspongia) sp. LG4, Cape Wellington, Wilsons Promontory. Mark Norman

Callyspongia (Callyspongia) sp. LG5, Cape Howe. Mark Norman

Subgenus *Callyspongia (Toxochalina)*

Defining characters of this subgenus are evident in the outer layer where there are three distinct mesh sizes and conspicuous conules on the surface. These sponges contain toxas as well as oxeas.

Callyspongia (Toxochalina) sp. LG1, Symonds Channel sponge gardens, Port Phillip Bay. Mark Norman

Genus *Dactylia*

These sponges typically have a plant-like growth form and contain no spicules in their skeletons. Their fibres are cored and strengthened instead by foreign material and sand. They have a smooth, sandy appearance as seen here in the close-up image. Despite the amount of sand and debris incorporated into their skeletons, they retain an elastic texture due to the fibre content. Species of *Dactylia* may be found in the coastal waters of most states of Australia and in New Zealand.

Dactylia sp. LG1, Nepean Wall, Port Phillip Bay. Mark Norman

Family Phloeodictyidae

Sponges in this family are differentiated from other haplosclerid sponges by having a detachable surface layer or crust. The arrangement of spicules in the skeleton of these sponges is termed **isotropic** (a mesh of single spicules without differentiation into a hierarchy of fibres). Phloeodictyid sponges are commonly burrowing in habit and often adopt a fistulose growth form of tubes that project above the substrate, as seen in the orange sponge *Oceanapia* sp. LG1.

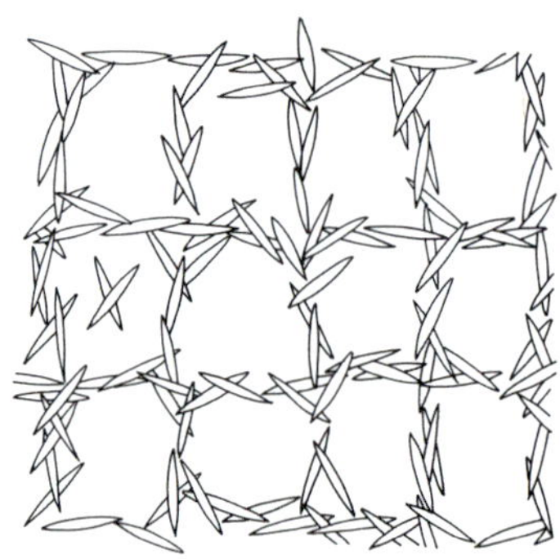

Isotropic skeleton

Genus *Oceanapia*

Sponges in this genus have a crust of spicules aligned tangential to the surface of the sponge. They often have a hollow 'body' with protruding fistules but may be globular or erect and lamellate. The mesh formed by the network of spicule tracts tends to be oval-shaped. Both of the species included here are fistulose in habit.

Oceanapia sp. LG1, Port Phillip Heads. Bay City Divers Group

Oceanapia sp. LG1, Port Phillip Heads. Ray Hill

Oceanapia sp. LG2, Refuge Cove, Wilsons Promontory. Mark Norman

Suborder Spongillina

Freshwater sponges all belong to this suborder. There are seven families and 45 genera of **extant** (living) freshwater sponges. It is thought that sponges spread from the oceans to inland waters somewhere between 140 and 210 million years ago, during the Jurassic Period. The success of their colonisation was due in part to the production of **gemmules** made up of a collection of totipotent cells, capable of regeneration of the 'mother' sponge. Sponges are now found throughout the world in freshwater lakes, rivers, estuaries and even some garden pools. They inhabit freshwater bodies from the shallowest of pools to depths over 100 metres, from the tropics to desert environments, and from the Arctic to as far south as Patagonia.

Most, *but not all*, freshwater sponges produce gemmules during periods of harsh conditions. Keys used to separate them to family and genus levels are reliant on the presence of gemmules in those sponges that *do* produce them.

Representatives of two families and 11 genera may be found in Australia. Most of them have a cosmopolitan distribution. One of the more commonly encountered species is included here and belongs in the family Spongillidae.

Freshwater sponge. Mark Norman

Family Spongillidae

This family contains exclusively freshwater sponges, exhibiting the full range of typical sponge growth habits, from encrusting on rocks and submerged logs, to branching or massive and irregular in form. Colour ranges from white to green or brown, texture from soft and delicate to hard and stony, and surface features from smooth to hispid or conulose. The skeleton is made up of a reticulate mesh formed by tracts of spicules, either oxeas or strongyles, which may be spiny or roughened in appearance.

Genus *Ephydatia*

Sponges in this genus are typically encrusting, bulbous or massive in habit. Spongin fibres are not commonly seen in the skeletons of these sponges, and the megascleres are oxeas which may appear smooth or roughened by **microspines**. The surface is velvety and the green colouration, if present, is due to the presence of **zoochlorellae** (symbiotic green algae).

Ephydatia fluviatilis was found at Darlot Creek, Lake Condah region in Western Victoria.
Mark Norman

Order Dictyoceratida

This order is one of the three orders of **collagenous sponges** that do not contain native spicules. All dictyoceratid sponges are supported by a spongin fibre skeleton. In most the fibres are organised as a hierarchy of larger primary, smaller secondary and, in some species, fine tertiary fibres. These sponges are generally tough and flexible, but can become hard and brittle due to their tendency to incorporate detritus into the matrix. If present, spicules found in these sponges are a foreign element of the incorporated debris. The taxonomy of these sponges is based on surface features and fibre characteristics. Dictyoceratid sponges often exhibit a darker outer region with a paler interior. There are four families in this order, all of which are represented here.

Family Irciniidae

The commonly occurring sponges of this widespread family are nearly always massive in habit and characteristically tough to cut or tear due to the presence of very fine collagen filaments. These are microscopically identifiable by an **ampule** (terminal swelling). Three genera are recognised in this family. Defining characters include the presence or absence of a sand-armoured crust (cortex) and the presence or absence of sand and detritus coring skeletal fibres.

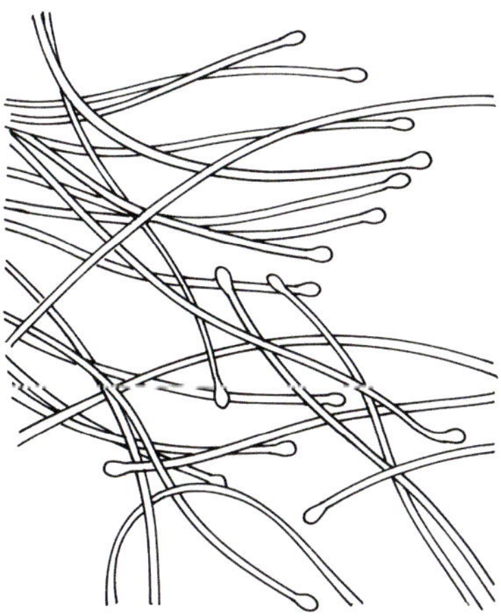

Collagen filaments with terminal ampules characterise sponges of the family Irciniidae.

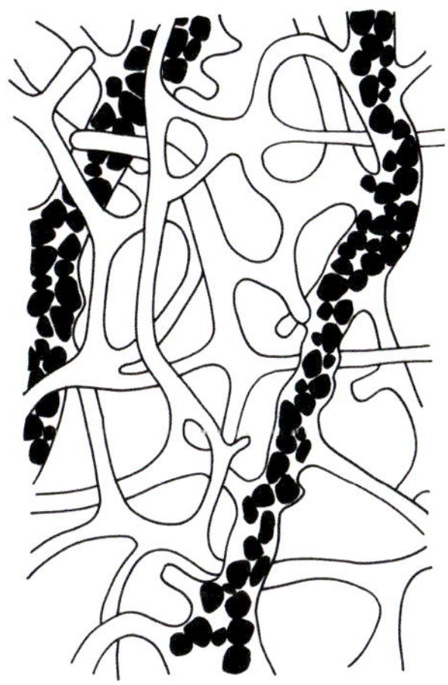

Fibres with a core of sand and detritus.

Ircinia sp. LG1, Port Phillip Bay. Mark Norman

Genus *Ircinia*

Sponges in this large genus are generally massive or caliculate but may be thickly encrusting, lobate or digitate. The surface is usually conulose. They are characterised by the *absence* of a sandy cortex, and by the *presence* of sand and detritus coring their fibres. The primary fibres of the skeleton often **anastomose** (fuse together) to form **fascicles** (bundles), giving these sponges the texture of 'old boots'.

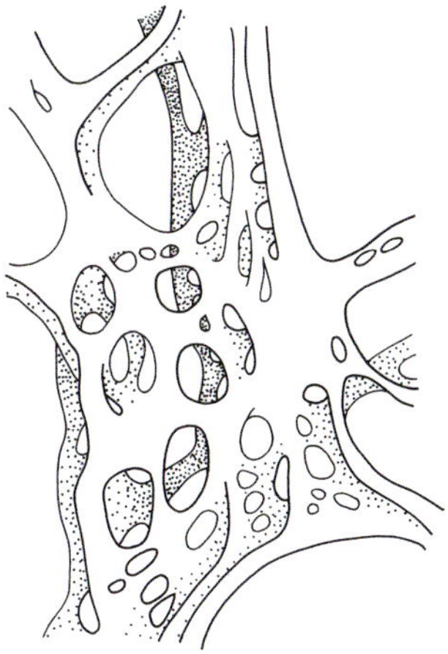

Primary fibre fascicle

Ircinia sp. LG2, Mornington Pier, Port Phillip Bay. Mark Norman

Ircinia sp. LG5, Port Phillip Bay. Mark Norman

Genus *Sarcotragus*

Sponges in this genus also lack the sandy cortex but the fibres are *clear* of coring sand and debris. Similarly, as in species of *Ircinia*, the primary fibres form fascicles. While often having the conulose surface feature, species of *Sarcotragus* tend to adopt a low and compact habit.

Sarcotragus sp. LG1, Mornington Pier, Port Phillip Bay. Mark Norman

Genus *Psammocinia*

These sponges have a greater variety of growth forms from compact and encrusting, erect and digitate to lamellate or vase-shaped. They are distinguished by the *presence* of a sand-armoured cortex and skeletal fibres that are *usually* cored by foreign material. Always firm and often incompressible in texture, sponges in this genus may be brittle due to the large quantities of sand incorporated into the skeleton.

Psammocinia sp.LG1 encrusting Steve's Bommie, Pillar Point, Wilsons Promontory. Julian Finn

Psammocinia sp. LG3, 'The Links', Lonsdale Wall, Port Phillip Bay. Mark Norman

Family Thorectidae

Representatives of this widespread family occur in all oceans except polar seas. The 23 genera are distinguished from other dictyoceratid sponges by their **laminated** (layered) fibres, appearing microscopically like layers of bark, and by a lack of the fine, collagen fibrils found in the Family Irciniidae. Taxonomy within the family is based on the presence or absence of a sand-armoured cortex, attributes of the fibre skeleton and, for some genera, the characteristic growth forms (*i.e.* honeycombed or digitate).

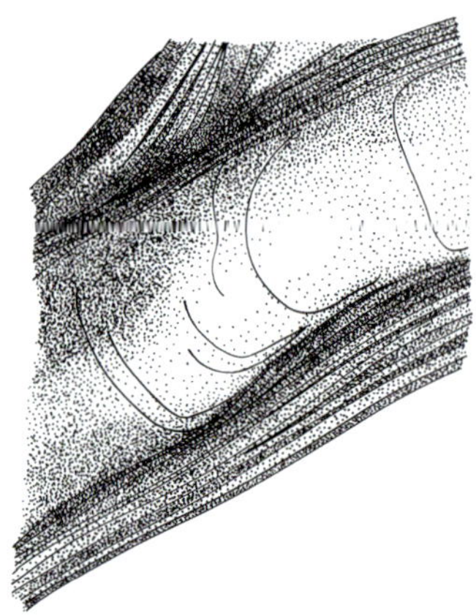

Concentric layers of a laminated fibre

Genus *Hyrtios*

Sponges in this genus are erect rather than encrusting and are generally tubular or digitate. The primary and secondary fibres of these sponges are completely cored or packed with sand and detritus, making it difficult to determine fibre structure under the microscope. These sponges have a roughened, conulose surface appearance, formed by the protruding, cone-shaped tips of fibres at the surface, as seen in the close-up below.

Hyrtios sp. LG1 (Mark Norman) with close-up inset (Julian Finn), 'The Links', Lonsdale Wall, Port Phillip Bay.

Genus *Taonura*

Sponges in this genus are characteristically soft and may be stalked, caliculate or **foliaceous** (branching with fronds), like the species shown here. They lack a sand-armoured cortex and have a regular network of fibres forming an almost rectangular mesh. The large primary fibres have a core of sand grains, while the connecting secondary fibres are uncored, or clear of sand.

Taonura sp. LG1, Symonds Channel sponge garden, Port Phillip Bay. Mark Norman

Genus *Thorectandra*

A number of features of this genus make its members more readily recognisable. They are always stalked and may be tubular, globular or flabellate. The large oscules are always **apical** (on the top of the sponge) or **marginal** (along the margins). The surface in some is characterised by prominent ridges and hollows, giving an almost honeycombed appearance. The cortex is armoured due to incorporation of sand into the outer layer of the skeleton. The texture, however, is quite soft and crumbly due to the wide-spaced mesh of fibres. These sponges typically exude copious amounts of thick mucus when handled. Most records of sponges in this genus are from Australia and New Zealand.

Thorectandra sp. LG3, Nepean Wall, Port Phillip Bay. Mark Norman

Thorectandra sp. LG3, close-up of sponge on previous page. Julian Finn

Thorectandra sp. LG3, Nepean Bay, Port Phillip Bay. Mark Norman

Genus *Fenestraspongia*

This genus takes its name from the characteristic **fenestrate** (grid-like) sponge surfaces, formed by intersecting ridges. The sponges in this genus are erect, thick and lamellate, or cylindrical-tubular in habit. They are also compressible but tough. Their skeletons are made up of a clearly hierarchical network of large primary, intermediate secondary and fine tertiary fibres. Only two species of *Fenestraspongia* have been described to date, both of which are found in south-eastern Australia and only one of which has been recorded from Victorian waters. The species included here is most likely that same species, *Fenestraspongia intertexta* (Carter, 1885).

Fenestraspongia cf. *intertexta*, Port Phillip Bay. Mark Norman

Genus *Carteriospongia*

These sponges tend to be lamellate, caliculate, **foliose** (leaf-like) or spreading over the substrate. The surface is characteristically ridged and appears sandy, due to the amount of incorporated sand and debris in the ectosome. The fibre skeleton again consists of primary, secondary and tertiary fibres. The larger, primary fibres may be filled with sand and may coalesce to form fascicles. The tertiary fibres are **vermiform** (long, thin and tangled like worms). These features combine to give these sponges a harsh or coarse texture, while leaving them flexible.

Carteriospongia sp. LG1 (Mark Norman) with close-up inset (Julian Finn), Lonsdale Wall, Port Phillip Bay.

Family Spongiidae

The familiar cosmetic or bath sponges belong in this family. All members are characterised by their **homogeneous** (unlaminated) skeletal fibre construction lacking a **pith** (central core). The six genera in this family contain no native spicules to aid identification. Instead they are distinguished using a combination of surface features and differences in the structure of their fibre skeletons. It is due to the density of the fibres in these sponges that they are both spongy and tough.

Genus *Hippospongia*

Almost completely lacking primary fibres, these sponges have a well-developed network of secondary and tertiary fibres, the tips of which form 'brushes' where they emerge at the surface. Internally these sponges have **lacunae** (holes) formed by the exhalant oscular cavities extending deep into the sponge 'body'. In life these sponges are darkly pigmented on the outer surfaces. Species of *Hippospongia* have been recorded from waters all around Australia. With little or no incorporation of sand into their skeleton, they are one of the few genera of sponges that are soft and elastic as well as tough. Once properly prepared, they are ideal for cosmetic use on human skin.

Hippospongia sp. LG1, Pillar Point, Wilsons Promontory. Julian Finn

Genus *Leiosella*

Sponges in this genus may be caliculate, flabellate, lobate or spreading over the substrate. All species are armoured with a thin layer of sand incorporated into the ectosome. Primary and secondary fibres make up the skeleton but it is the dense secondary fibre network that characterises these sponges and leaves them compressible yet firm – not suitable for bathroom use. Representatives of the genus occur in coastal waters of all Australian states.

Leiosella sp. LG2, Nepean Bay, Port Phillip Bay. Mark Norman

Family Dysideidae

These commonly occurring sponges are found in shallow, coastal waters from the tropics to temperate seas. They tend to be soft with a conulose surface. Microscopically, the fibres in the skeleton show both lamination and pith, although these features may be difficult to see in two of the five genera that incorporate sand and debris into their skeletons. The texture may become harsh or brittle depending on the amount of incorporated debris.

Genus *Euryspongia*

These dysideid sponges incorporate sand and debris into the primary fibres of their skeleton, while secondary and tertiary fibres are clear of detritus. They tend to be soft and collapsible and, if large quantities of detritus are incorporated, they can become quite delicate and fragile. They are most commonly encrusting over the substrate, although they can be massive or branching. The conulose surface feature is present but not always prominent.

Euryspongia sp. LG3 close-up, Cape Wellington, Wilsons Promontory. Julian Finn

Order Dendroceratida

Sponges of this collagenous order also lack spicules. They are often soft and very fragile due to a reduction in the fibrous skeleton in relation to the tissue mass. All begin their growth from a base that spreads over the substrate. The fibres of the skeleton then grow either in a branching, **dendritic** pattern (family Darwinellidae) or a mesh-like growth pattern (family Dictyodendrillidae). The fibres of all sponges in this order show strong concentric lamination and a pronounced pith, visible microscopically as a dark, opaque central line of collagen within the fibre. Examples of both families occur in southern Australian waters but only one family is included here.

Family Darwinellidae

The fibre skeleton of sponges in this family is always dendritic, never reticulate. Darwinellid sponges are generally encrusting in habit but can be erect and branching. Of the four genera in this family, three are included here.

Genus *Darwinella*

Sponges in this cosmopolitan genus are found from temperate to tropical waters at depths down to 200 metres. They are characterised by having a fibrous skeleton devoid of sand or detritus, supplemented by collagenous spicules (as opposed to mineral-based siliceous or calcareous spicules). These fibrous spicules are found free in the collagen matrix. They may have two, three or more axes with tapered points and tend to be curved (in contrast to straighter mineral-based spicules). Fibrous spicules appear the same density and colour as the fibres themselves when viewed through a microscope. When distinguishing between species of *Darwinella* and *Aplysilla*, easily confused, the presence or absence of fibrous spicules is an important microscopic feature to consider.

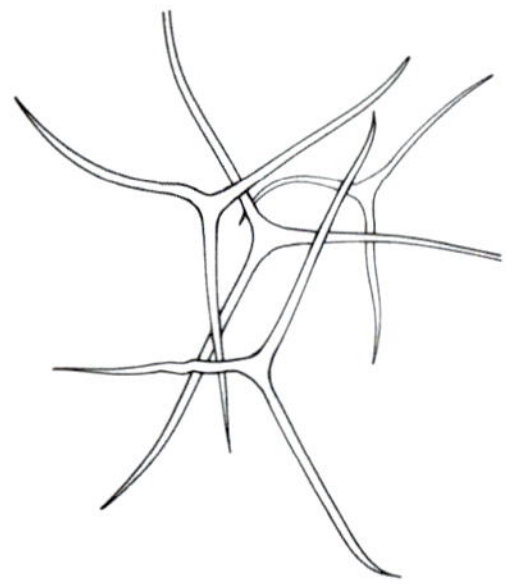

Fibrous sponge spicules

Typically soft and fleshy, species of *Darwinella* range in shape from encrusting to massive or lobate. They are often yellow or pink-red in colour. If damaged or exposed to air, the yellow specimens typically turn dark purple. This colour change is termed **aerophobic**. The species *Darwinella australiensis* occurs commonly in Victorian coastal waters and Bass Strait.

A species of *Darwinella* encrusting the sea squirt *Pyura spinifera*. Mark Norman

Darwinella australiensis encrusting a boulder, Mornington Pier, Port Philip Bay. Mark Norman

Genus *Aplysilla*

Species of *Aplysilla* occur worldwide in water ranging from very shallow to over 600 metres deep. Easily confused with other encrusting genera in this family, species of *Aplysilla* can really only be identified using microscopic features. Member species are found from the Great Barrier Reef in Queensland to Tasmania, commonly encrusting boulders and pier piles. *Aplysilla rosea*, shown here encrusting boulders, may also be found growing on scallop shells.

A sea slug nudibranch feeding on a sponge, *Aplysilla rosea*. Julian Finn

Dendrilla cactos encrusting form with inset showing amphipods, Pillar Point, Wilsons Promontory.
Julian Finn

Genus *Dendrilla*

Sponges of the genus *Dendrilla* are not common but can be found in oceans from Antarctica to the tropics, most often at depths of 10–50 metres. They adopt an erect, branching or lamellate growth form but, like all sponges of the Order Dendroceratida, they have an early encrusting growth phase and at this stage are easily confused with other encrusting genera in the order. The collagenous fibres of these sponges are dendritic. The surface is smooth between the prominent conules and may appear almost 'frilly' in some specimens. *Dendrilla cactos,* seen here in both its encrusting and erect branching forms, occurs in rocky reef habitats of southern Australia.

Dendrilla cactos with encrusting worms, Cape Wellington, Wilsons Promontory. Mark Norman

Order Verongida

This is the third of the three orders of collagenous sponges that contain no spicules. The skeletons of these sponges are entirely fibrous. The fibres appear characteristically dark microscopically and are laminated like the layers of bark on a tree. They also contain a core of pith visible by microscopy as a dark line. There is no distinction in size between primary and secondary fibres and they only rarely contain **detritus** (foreign debris). The skeleton formed by the fibres may be either anastomosing or dendritic. While the skeletal fibres are quite tough they are flexible and the sponges generally feel quite fleshy and easily deformable. Verongid sponges are often pigmented yellow with a greenish tinge at the edges and are aerophobic becoming purple to black if damaged or exposed to air. They are commonly stalked and branching or vase-like, but may also be thin and encrusting. Four families make up the order, only one of which is presented here.

Family Aplysinidae

The skeletal fibres of these sponges form a polygonal-shaped mesh in one or more planes. All sponges in this family exhibit an aerophobic colour-change on death or when damaged, changing from the typical yellow-green to purple-black. Some of the chemicals produced as a by-product of metabolism by sponges in this family exhibit strong antimicrobial activity. One of the three genera in this family is included here.

Genus *Aplysina*

The skeleton of these sponges is a regular, rectangular mesh of uniform fibres, each of which contain a thick pith. All sponges in this genus exhibit a distinctive yellow colouration in life, and the aerophobic colour change on death or if damaged. The species *Aplysina lendenfeldi* is very common around the coast of Victoria and extends into New South Wales, South Australia and Tasmania. It typically adopts a characteristic stalked, cigar-shaped growth form with a single oscule on the upper surface as seen here.

Aplysina lendenfeldi with close-up below, Beware Reef, East Gippsland. Julian Finn

Class Calcarea

This class is restricted to marine sponges with skeletal elements (spicules and/or solid cemented base) being composed of calcium carbonate. They are found worldwide and are not uncommon but appear less obvious than the siliceous sponges (Class Demospongiae) because they tend to occupy more cryptic marine habitats such as caves and under overhangs. Calcareous sponges are generally smaller and less colourful than their relatives the demosponges.

The spicules of the Calcarea are completely different in nature from those found in the demosponges. They can have multiple axes of symmetry – two (**diactine**), three (triactine), four (tetractine) or more (**polyactine**).

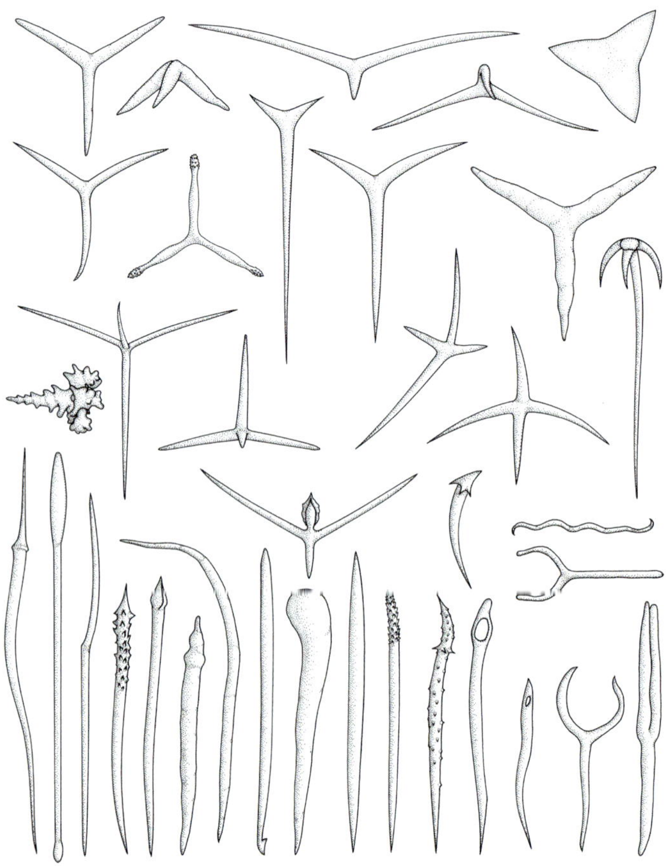

Spicules of the calcareous sponges.

Clathrinid sp. LG3, Wilsons Promontory. Mark Norman

In ancient seas calcareous sponges were an extensive and ecologically important group, responsible for much of the reef-building of the time. Extant calcareous sponges, however, account for only about 5% of described sponge species, due in part to a limited number of experts and a lack of taxonomic effort. At present the Class Calcarea is made up of two subclasses, five orders, 22 families and 75 genera.

Calcareous sponges are not readily identifiable one from another but, as for the demosponges, their taxonomy is based on a combination of characters. These include spicule morphology, the presence or absence of a cemented basal skeleton, features of the aquiferous system, larval development and cellular structure, most of which require specialist knowledge and laboratory procedures. Identification beyond the level of order is best achieved using molecular techniques. There are many calcareous sponges to be found in Victorian waters, a few representatives only are presented here.

Order Clathrinida

The skeletons of these sponges contain only *free* spicules having no rigid or cemented parts of the skeleton. The spicules may have three or four axes of equal length (**equiradiate**) set at equal angles (**equiangular**). This is an important feature used to classify calcareous sponges at this level. There are six families and 16 genera of Clathrinid sponges.

Clathrinid sp. LG1, Pillar Point, Wilsons Promontory, depth 10 m.
Julian Finn

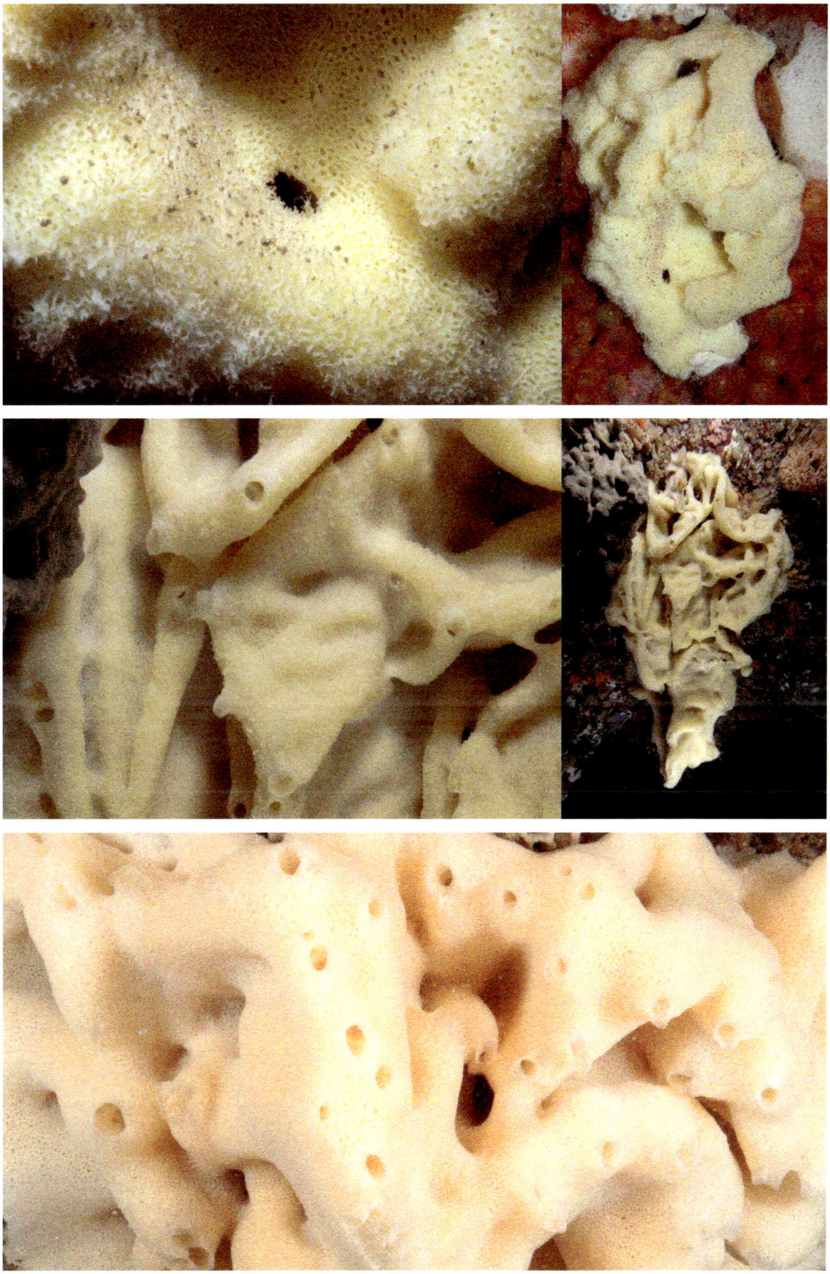

Clathrinid sp. LG2 – top (Julian Finn) and Clathrinid sp. LG3 –middle (Mark Norman), Wilsons Promontory. Clathrinid sp. LG4 – bottom (Mark Norman), Port Phillip Bay.

Clathrinid sp. LG5, Port Phillip Bay. Mark Norman

Clathrinid sp. LG5 close-up, Port Phillip Bay. Mark Norman

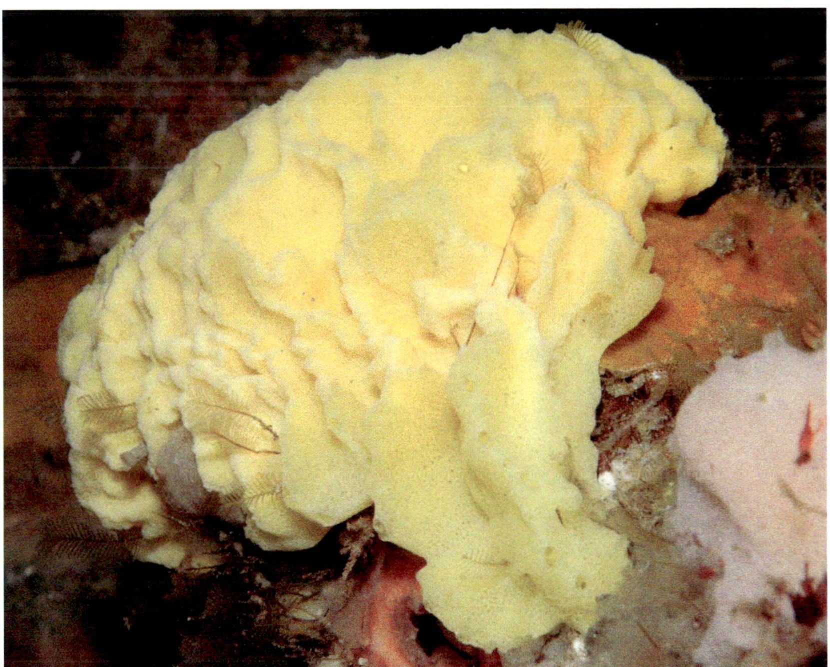

Clathrinid sp. LG6, Pumphouse Bay, Gabo Island, depth 10 m. Mark Norman

Sycon sp., St Leonards, Port Phillip Bay. Mark Norman

Order Leucosolenida

As for the clathrinid sponges, only *free* spicules (*i.e.* not fixed or cemented) are found in the skeletons of leucosolenids. The spicule morphologies in sponges of this order differ however, and may include diactines, and tri- or tetractines, having three or four axes set at angles that are *unequal*. This is an important feature used to classify calcareous sponges at this level. At present there are nine families and 42 genera of sponges in this order. One of the more common examples is featured here.

Family Sycettidae

Calcareous sponges in this family consist of a central tube surrounded by a radial arrangement of secondary tubes growing perpendicularly. All internal surfaces of the tubes are lined by a layer of current-producing choanocyte cells. At the ends of the tubes, tufts of long, diactine spicules are visible by eye.

Genus Sycon

Members of this cosmopolitan genus are commonly seen growing in clusters of small tube-like fingers, usually attached to hard substrates but occasionally in soft sediments.

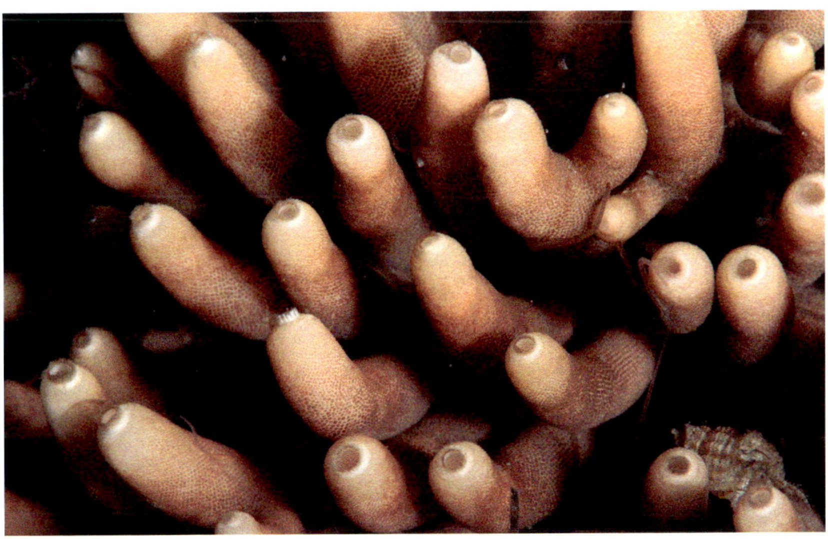

Sycon sp., close-up of sponge opposite. Mark Norman

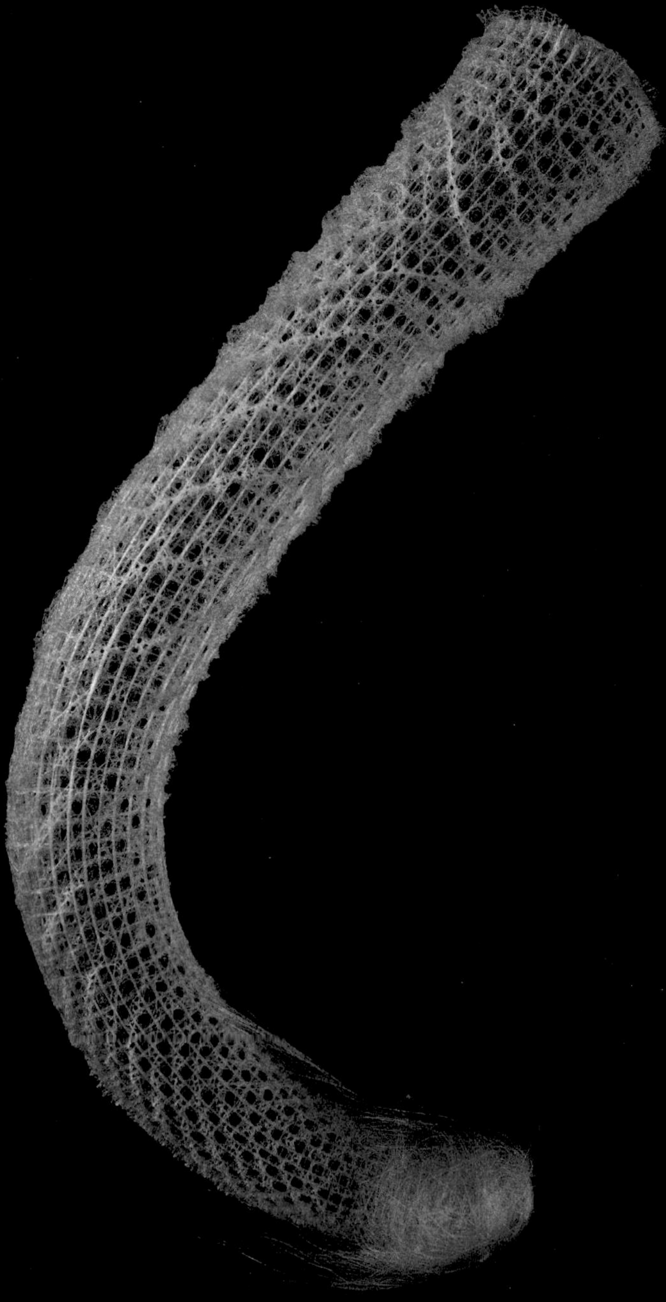

Glass sponge, *Euplectella aspergillum* (detail, right), from the Museum Victoria collection. David Paul

Class Hexactinellida

Class Hexactinellida, the glass sponges, includes marine sponges with spicules made of silicon dioxide, the substance used in glass manufacture. Hexactinellid spicules are typically six-rayed (**hexactine**). They have been recorded from depths between 5 and 7000 metres but account for only about 7% of all described living sponge species. In deep-sea environments, however, glass sponges are important members of benthic communities.

Growth forms of the glass sponges include the wide range of shapes found in the demosponges, such as tubular, caliculate, lobate, branching and more massive forms, however they lack encrusting forms. Unlike the skeletons of the previous two classes, the skeletons of glass sponges tend to consist of a lattice-like construction of separate or fused spicules, over which lies a thin layer of 'tissue' formed by cells that have coalesced into a multinucleate mass or **syncitium**. The collar bodies producing the water flow of glass sponges are different too. They are not the choanocyte cells of the previous sponge classes but rather units possessing single beating flagella that line the chambers. Glass sponges contain some collagenous material but no true spongin fibres.

The spicules found in hexactinellids show a wide variety, some of which have very specific roles in the anchoring of sponges in soft sediment, or for attachment onto hard substrates.

Spicules of the class Hexactinellida

Sponges of the class Hexactinellida are separated into two sub-classes (based on microsclere morphology) and at present are further subdivided into five orders, 17 families and 118 genera. Examples across the five orders have been found from Australian waters, however there are no records yet from coastal Victoria.

APPENDIX

HISTOLOGICAL SLIDE PREPARATION TO AID SPONGE IDENTIFICATION

Note: Slide preparation should only be conducted by specialists in an appropriately equipped laboratory with adherence to strict safety procedures and guidelines.

SPICULE PREPARATION

A small fragment of sponge, approximately 1 mm^3, including both surface ectosome and inner choanosome, is placed directly onto a microscope slide. A drop of nitric acid is added to the sponge tissue and gently warmed over a small flame. A bubbling reaction takes place and all organic material is dissolved, leaving only siliceous spicules (as well as any sand grains) on the slide. Once cooled a few drops of mounting medium (e.g. Canada Balsam or Euparal) can be added and then a coverslip applied.

NB: Care must be taken when using nitric acid as it is highly caustic and will 'burn' skin. Safety glasses must be worn.

An alternative method of separating out spicules, especially those suspected of being calcareous, is to use household bleach. The fragment of sponge is placed into a small vial containing 10–20 ml of bleach. This is left for a period of at least 12 hours and then 'washed' by pipetting off the bleach and replacing it with water. The washing process is repeated two to three times and finally 70% ethanol is added to the vial. Allow the spicules up to 10 minutes to settle each time or they will be lost in the decanting process. Spicules may then be drawn up by pipette and placed directly onto a microscope slide. The ethanol will evaporate, leaving behind dry spicules. Again a few drops of mounting medium are placed on the spicules and a coverslip added.

SKELETAL SECTION PREPARATION

Various techniques (most of them requiring specialist equipment and harmful chemicals) are available for preparing a thin section of sponge tissue in order to view spicules and fibrous elements in place in the sponge skeleton. The simplest technique is hand-sectioning, using a scalpel and cutting against a straight edge (such as a microscope slide), to create the thinnest slices possible by eye. The slices should be made through a block of sponge approximately 5 mm^3 and care taken to include both ectosome and choanosome. Three or four of the thin sections can then be placed onto a microscope slide and covered with a coverslip (weighted with a small weight to keep the sections flat while drying). Air-drying will take 24–48 hours, preferably at a temperature of 30–40°C. A mountant such as Euparal can then be added to the dried sections and the coverslip reapplied.

Full instructions may be found in '*Sponguide: Guide to Sponge Collection and Identification*' (Version 2003), available on the Queensland Museum website.

FURTHER INFORMATION

This guide has been written as a practical introduction to southern Australian sponges. Those keen to learn more can consult the publications and websites listed below. Please note this is only a small selection of many important texts on sponge biology and taxonomy.

PUBLICATIONS

Much has been learned since the writing of this book but it remains a comprehensive text on the biology of sponges:
Bergquist, P.R. 1978. *Sponges.* (Hutchinson: London; and University of California Press: Berkeley and Los Angeles).

A thesaurus of terminology relating to the biology and taxonomy of sponges:
Boury-Esnault, N. and Rutzler, K. (Eds). 1997. *Thesaurus of Sponge Morphology. Smithsonian Contributions to Zoology,* 596: 1-55.

This atlas provides a detailed look at sponges at the cellular level through electron micrographs and accompanying diagrams:
De Vos, L., Rutzler, K., Boury-Esnault, N., Donadey, C. and Vacelet, J. 1991. *Atlas of Sponge Morphology.* (Smithsonian Institution Press: Washington and London).

For those determined to classify sponges, this two-volume, 2000-page key to the genera of sponges is an essential starting point:
Hooper, J.N.A. and Van Soest, R.W.M. (eds). 2002. *Systema Porifera. A Guide to the Classification of Sponges.* (Kluwer Academic / Plenum Publishers: New York).

A complete catalogue of the described sponge species of Australia as at the time of writing:
(More up to date listings available on the *Australian Faunal Directory* and *World Porifera Database* websites below.)
Hooper, J.N.A. and Wiedenmayer, F. 1994. Porifera. *In:* Wells, A. (ed). *Zoological Catalogue of Australia.* (Melbourne: CSIRO Australia). Volume 12: 1-624.

A personal account of the history of diving for sponges:
Kalafatas, M.N. 2003. *The Bellstone. The Greek Sponge Divers of the Aegean.* (Brandeis University Press: New England, USA): 1-287.

WEBSITES

***Australian Faunal Directory* – checklists of Australian species:**
http://www.environment.gov.au/biodiversity/abrs/online-resources/fauna/afd/groups

Introduction to the phylum Porifera:
http://www.ucmp.berkeley.edu/porifera/porifera.html

Museum Victoria, *Natural Sciences Online:* a selection of collection specimens including species identification and distribution data:
(http://collections.museumvictoria.com.au/search.php),

Queensland Museum *Sponguide:* Guide to sponge collection and identification (2003)
http://www.qm.qld.gov.au/Find+out+about/Animals+of+Queensland/Sea+Life/Sponges/Collecting+and+preserving+sponges

Queensland Museum website – the nature of sponges and sponge research in Queensland:
http://www.qm.qld.gov.au/Find+out+about/Animals+of+Queensland/Sea+Life/Sponges

Taxonomic toolkit for marine life of Port Phillip Bay. This site was funded by the Victorian Government Department of Sustainability and Environment *Seagrass and Reefs Program for Port Phillip Bay*. It provides information, data, images and tools to help identify, document and monitor the marine animal life of Port Phillip Bay.
http://portphillipmarinelife.net.au

***World Porifera Database* – worldwide checklist of sponges with original citations:**
http://www.marinespecies.org/porifera/

AUTHOR ACKNOWLEDGEMENTS

We thank the sponge community world-wide, upon whose work we have drawn, notably Patricia Bergquist, John Hooper, Rob Van Soest, and all of the contributing authors of *Systema Porifera*. We thank the staff of Parks Victoria, especially Roger Fenwick, Steffan Howe and Chris Hayward, for support in photographing and collecting sponges for this guide. We thank Jeanette Birtles for editorial supervision, Dave Collins for coordinating illustrations, and particularly thank Melanie Raymond and the production team at Museum Victoria.

PHOTOGRAPH AND ILLUSTRATION CREDITS

We thank those who provided additional images, including:

Bay City Divers Club, Tim Allen, Ron Blakely, John Chuk, Kerry B Clark, Peter Clarkson, Amanda Coakes, Neville Coleman, Mark Davel, Ray Hill, Manfred Krautter, Peter Kyne, Sally Leys, David Paul, Joseph R. Pawlik, Andrzej Pisera, Mark Rosenstein, T.G. Thompson, University of Queensland and Jean Vacelet.

We would especially like to thank Judith Lockhart for volunteering her time to draw the illustrations that feature throughout this guide and Leonie Hooper, as well as ABRS, for permission to reprint the spicule drawings.

ABOUT THE AUTHORS

Lisa Goudie

Lisa is a consultant sponge taxonomist and works for both government and private organisations. She has spent many years on research and fishing vessels collecting marine invertebrates for museum collections and data for fisheries. Lisa is currently self-employed providing identification of sponges for researchers in fields such as biodiversity, environmental impact studies and marine natural product discovery.

Dr Mark Norman

Dr Mark Norman is Head of Sciences at Museum Victoria where he leads the large and active natural sciences research team of curators, collection managers, postdoctoral fellows, postgraduate students, research associates and volunteers. Mark's primary research interest is cephalopods (octopuses, squids, cuttlefishes and nautiluses). Specific research interests include biodiversity, evolutionary origins and relationships, reproductive and defensive behaviours, deep-sea faunas and adaptations to key habitats and niches. Mark has published field guides on cephalopods and marine invertebrates and children's books on natural history. He has participated in numerous natural history documentaries.

Dr Julian Finn

Dr Julian Finn is a Senior Curator of Marine Invertebrates at Museum Victoria. He has years of experience in marine invertebrate research (primarily studying cephalopods) and is involved in diverse projects revealing and promoting Victoria's unique marine environment. Julian has discovered and described new species and genera, reported novel behaviours in octopuses and cuttlefishes and worked as a freelance cameraman and scientific consultant on over 30 international documentaries, shooting stories in diverse locations including Japan, Indonesia and Mexico. Julian's work has taken him from the tropics of Baja California to the chilly waters of Antarctica, and his images feature in many of the exhibitions, marine field guides and publications produced by Museum Victoria.

GLOSSARY

Acanthostyle spined style

Actinocyte sponge cell type involved with contraction

Aerophobic colour change following damage or exposure to air

Amphiaster star-shaped **microsclere** with rays radiating from both ends

Amphiblastula first **larval** stage of a sponge following fertilisation

Ampule terminal swelling on a **collagen fibril** found in the sponge family Irciniidae

Anastomosing fusing together to form a mesh-like skeleton of fibres

Anisochelae chelae with unequal ends

Apical positioned at the top

Aquiferous water-bearing

Arborescent tree-like

Archaeocyte cells capable of transforming into any other sponge cell type

Areolae groups of **inhalant pores** on the surface of some sponges, also known as **pore** sieves

Asconoid simplest sponge body form where the entire internal surface is lined by current-producing cells

Asexual without the union of egg and sperm

Asterose star-shaped **spicule** type

Axis of symmetry a plane dividing a body into two symmetrical halves (plural axes)

Bacteriocytes cells containing bacteria which provide amino acids and other chemicals to their host organism

Bellstone a flat stone used to send a diver more quickly to greater depths

Benthic living on the seafloor, bottom-dwelling

Birotule type of **chelate microsclere** with a circle of teeth at each end

Calcareous having calcium carbonate components

Calcareous sponges one of the three classes of sponges (Calcarea) – marine sponges with skeletal elements composed of calcium carbonate

Caliculate cup-shaped

Chelae/chelate c-shaped **microscleres** peculiar to sponges in the order Poecilosclerida

Choanocyte cell type which produces the current flow through sponges and traps food particles

Choanoderm the innermost layer of cells (**choanocytes**) lining the canals and chambers of sponges, responsible for generating water currents

Choanosome internal region of a sponge

Choanoflagellate single-celled organism with a beating **flagellum** and collar of **cilia** like the **choanocyte** cell of sponges

Cilia hair-like structures on the **choanocyte** cell that trap tiny food particles

Collagen fibrous protein found in the **mesohyl** of sponges

Collagenous sponges three orders of sponges that do not contain native spicules

Commata comma-shaped **microscleres**

Conules cone-shaped projections on the sponge surface raised up by the skeleton beneath, usually visible with the naked eye

Conulose having cone-shaped projections on the surface

Coring forming the core of a fibre (with **spicules**, sand and/or **detritus**)

Cortex outermost layer

Cortical of the **cortex**

Demosponges one of the three classes of sponges, contains 85% of living species (including most marine and all of the freshwater sponges)

Dendritic branching, tree-like fibre skeleton

Detritus foreign debris

Diactine having two **axes of symmetry**

Digitate finger-like

Echinating spicules embedded in and protruding perpendicular to sponge fibres

Ectosome outermost, superficial region of a sponge, distinct from the inner **choanosome**

Elongate drawn-out form

Endemicity restriction of the natural distribution (of a particular species) to a particular region

Equiangular with all angles equal

Equiradiate axes of equal length

Exhalant outward-moving passage of the water current

Extant still existing

Fascicles bundles of coalesced primary fibres found in some sponges

Fenestrate grid-like pattern of ridges on the surface of sponges of the genus *Fenestraspongia*

Fibrils tiny **collagenous** units found free in the mesoderm or in the **pith** of some fibres

Fistules finger-like hollow tubes which project from the surface of some excavating sponges, often terminating in **oscules**

Fistulose a growth form with hollow tubes called **fistules** projecting from the surface

Fixed preserved

Flabellate/Flabelliform fan-shaped

Flagellum thread-like filament of a choanocyte cell, capable of movement and responsible for creating the water flow through sponges

Foliaceous branching with fronds

Foliose leaf-like

Foreign (spicules) not produced by the organism containing them

Gemmule a concentrated mass of cells capable of withstanding harsh or dry conditions before regenerating

Genus taxonomic grouping of closely related species

Habit growth form

Halichondroid apparently haphazard arrangement of skeletal elements

Hexactine six-rayed **spicules** found in the Class Hexactinellida

Hexactinellids one of the three classes of sponges (known as 'glass sponges')

Hierarchy distinct levels of fibre size

Hispid velvety appearance

Homogeneous unlaminated fibres as seen in sponges of the Family Spongiidae

Inhalant for inward-moving passage of the water current

In situ in the natural location

Inorganic not **organic**

Invertebrate an animal without a vertebral column (backbone)

Isodictyal triangular mesh of skeletal elements the length of a single **spicule**

Isotropic mesh of single **spicules** without differentiation into a hierarchy of fibres

Lacunae holes or cavities within the sponge body

Lamellate plate-like

Lamination layered formation of skeletal fibres

Larva immature form of an invertebrate animal (plural larvae)

Leuconoid most evolved and complex body plan of sponges with folding of the canals to form chambers

Lobate having lobes

Marginal position of features (*e.g.* **oscules**) along the margins or edges

Massive large, without a definable shape

Megasclere large, structural **spicule**

Mesohyl middle region of a sponge containing skeletal and cellular elements

Microrhabd short, spined, rod-like **microsclere**

Microsclere small, reinforcing spicule, often of unique shape

Microspines microscopic spines on a **spicule** (**megasclere** or **microsclere**) giving it a roughened appearance

Microstrongyles rod-shaped **microsclere** rounded at both ends

Microxeas needle-like **microscleres** pointed at both ends

Monaxone spicule with a single axis

Morphologies structural forms

Motility ability to move

Mycalostyles rod-shaped **megascleres** that have a slight constriction at the neck of the rounded end, characteristic of the sponge family Mycalidae

Native refers to the spicule content of sponges *i.e.* those produced by the **sclerocyte** cells of the sponge itself

Octocoral subclass of **corals** with eight-fold symmetry

Onychaetes small, spined, rod-shaped **microscleres**

Organic a group of molecules containing carbon and hydrogen, defining life

Oscules exhalant apertures of a sponge

Ostia inhalant pores of a sponge

Oxea rod-shaped **megasclere** pointed at both ends

Palisade a continuous brush of **spicule** tips at the sponge surface

Pedicels root-like structures anchoring a sponge to the sea floor (*e.g. Tethya*)

Pinacocyte sponge cell type forming the outer layer

Pinacoderm outermost layer of cells of a sponge

Pith fine fibrillar material visible in the centre or core of some fibres

Plasticity changeability

Polyactine many **axes of symmetry**

Pore opening to the surface through which water currents move

Porifera sponges, Latin for pore-bearers

Primary largest fibre size

Propagule asexual bud (*e.g.* produced by sponges of the genus *Tethya*)

Radula rasping toothed tongue of a mollusc

Raphide hair-like microsclere

Raspailiid surface feature of sponges in the family Raspailiidae; a single, large protruding **spicule** surrounded by a bouquet of smaller spicules at the base

Reticulate having a mesh, or network, of fibres in the skeleton

Scaphander diving suit of the 1800s

Sclerocyte cell which secretes **inorganic spicules**

Secondary second-largest fibre size

Sequential hermaphrodites individuals which produce eggs, then later produce sperm (or *vice versa*), thus avoiding self-fertilisation

Sigma c- or s-shaped microsclere

Sinuous curved or snake-like

Spicular/Spiculose containing **spicules**

Spicules inorganic glass-like microscopic structures made of silica or calcium carbonate

Spirasters spiral, rod-shaped spiny **spicules**

Spongin collagenous substance binding **spicules** or forming fibres in the skeleton of sponges

Strongyle needle-like **megasclere** rounded at both ends

Style needle-like **megasclere** pointed one end, rounded the other

Stylote containing **styles**

Subtylostyle rod-shaped **megasclere** pointed one end with a subterminal bulb at the other

Syconoid sponge body plan with some folding of the **choanocyte** cell layer

Symbiotic two species living in relationship with one another and deriving mutual benefit from the association

Syncitium multinucleate mass of cellular material, as found in the Class Hexactinellida

Tangential skeletal elements lying parallel to the surface of the sponge

Tertiary third largest fibre size

Tetractine four-rayed **spicule**

Tetraxone spicule with four **axes of symmetry**

Totipotent the ability of a cell to transform to a different cell-type

Toxa a small, curved or wave-like **microsclere**

Tract column of **megascleres**

Triactine three-rayed **spicule**

Triaene a **tetractinal megasclere**, having one long ray and three short rays

Triaxone spicule with three **axes of symmetry** at right angles making six rays

Tubercle small rounded process projecting from the surface

Tuberculose having a surface covered in **tubercles**

Tylostyle pin-shaped **megasclere**

Tylote megasclere, pin-head shaped at both ends

Undifferentiated uniform in appearance, not specialised into different types

Unguiferous anisochelae asymmetrical chelae microscleres that have toothlike projections at each end

Vermiform tangled, worm-like pattern of tertiary fibres

Vestigial a character that is seen but reduced in size or prevalence

Voucher a specimen kept, usually by a museum, as a reference

Zoochlorella a green alga living symbiotically within an invertebrate animal (*i.e.* both the alga and the host gain from the relationship)

Zoophytes 'plant-animals' (an outmoded classification)

INDEX TO SCIENTIFIC NAMES